THE VOLKSWAGEN SUPER BEETLE HANDBOOK

From the Editors of *VWTrends* magazine

HPBOOKS

HP BOOKS
An imprint of Penguin Random House LLC
375 Hudson Street
New York, New York 10014
penguin.com

First edition: January 2006

Copyright © 2005 by Primedia, Inc.

Penguin supports copyright. Copyright fuels creativity, encourages diverse voices, promotes free speech, and creates a vibrant culture. Thank you for buying an authorized edition of this book and for complying with copyright laws by not reproducing, scanning, or distributing any part of it in any form without permission. You are supporting writers and allowing Penguin to continue to publish books for every reader.

Book design and production by Michael Lutfy

Cover design by Bird Studios

Cover photos Rob Hallstrom, courtesy *VWTrends* magazine archives

ISBN 978-1-55788-483-1

Printed in the United States of America

11

The information in this book is true and complete to the best of our knowledge. All recommendations on parts and procedures are made without any guarantees on the part of the author or the publisher. Tampering with, altering, modifying or removing any emissions-control device is a violation of federal law. Author and publisher disclaim all liability incurred in connection with the use of this information. The information contained herein was originally published in *VWTrends* magazine and is reprinted under license with PRIMEDIA Specialty Group, Inc., a PRIMEDIA™ company and © 2006 PRIMEDIA Specialty Group, Inc. All rights reserved.

CONTENTS

Sources — iv
Introduction — v

Chapter 1: The Rise and Fall of the Volkswagen Super Beetle — 1
Chapter 2: Initial Inspection and Evaluation — 10
Chapter 3: Interior/Exterior Disassembly — 17
Chapter 4: Front Suspension Disassembly — 25
Chapter 5: Front Suspension Rebuild — 33
Chapter 6: Rebuilding the Steering Box — 39
Chapter 7: Replacing The Package Tray — 43
Chapter 8: Rebuilding the Brakes — 48
Chapter 9: Replacing the Spring Bushing Plate — 53
Chapter 10: Removing Undercoating — 58
Chapter 11: Basic Bodywork — 64
Chapter 12: Removing Window Glass & Door Locks — 69
Chapter 13: Sanding & Priming — 74
Chapter 14: Removing the Body from the Pan — 78
Chapter 15: Replacing Rusted Fenderwells — 83
Chapter 16: Final Paint — 87
Chapter 17: Assembling the Long Block — 89
Chapter 18: Replacing the IRS Transmission — 101
Chapter 19: Installing the Headliner — 111
Chapter 20: Installing Rubber, Doors and Windows — 117
Chapter 21: Installing Wiring Harness and Electrical Components — 124
Chapter 22: Soundproofing Our Beetle — 131
Chapter 23: The Fuel and Fresh Air Vent Systems — 135
Chapter 24: Finishing the Interior — 142
Chapter 25: Installing Exterior Body Parts and Trim — 157
Technical Specifications — 168

SOURCES

A-1 Muffler
721 S. Main Street
Santa Ana, CA 92701
(714) 836-7201

Coker Tires
13187 Chestnut Street
Chattanooga, TN 37402
(423) 265-6368

Classic VW Specialty
Rafael Gutierrez
Orange, CA
(714) 778-0503

Dynamic Control
3042 Symmes Road
Hamilton, Ohio 45015
(800) 225-8133
www.dynamat.com

Engineered Application
4727 East 49th Street
Vernon, CA 90058
(323) 585-2894

Eastwood
263 Shoemaker Rd.
Pottstown, PA 19464
(800) 345-1178
www.eastwoodcompany.com

G&M Schapp
12520 Magnolia Ave., Unit L
Riverside, CA 92503
(951) 734-6382

Mid America Motorworks
One Mid America Place
P.O. Box 1248
Effingham, IL 62401
(800) 588-2844
www.mamotorworks.com

R&R Sandblasting
12520 Magnolia Ave., Unit K
Riverside, CA 92503
(951) 738-1516

Strictly Foreign
6760 N. Applegate Road
Grants Pass, OR 97527
(541) 862-7015
www.strictlyforeign.com

Top Line Parts
2910-A Miraloma Ave
Anaheim, CA 92806-1811
(714) 630-8371
www.toplineparts.com

Totally Stainless
P.O. Box 3249
Gettysburg, PA 17325
(717) 677-8811
www.totallystainless.net

West Coast Classic VW
1002 E. Walnut
Fullerton, CA 92831
(714) 871-1322
www.classicvw.com

West Coast Metric
24002 Frampton Ave.
Harbor City, California
(800) 847-3202
www.westcoastmetric.com

Wolfsburg West
2850 Palisades Dr.
Corona, CA 92880
(909) 549-0525
www.wolfsburgwest.com

INTRODUCTION

During my nearly six years as editor of *VW Trends* magazine, not many people I met admitted that they liked Super Beetles. I've heard people call them bulbous, awkward, bulky, swollen, sluggish, fat (no, not phat!)... someone even called it preposterous. In fact, in a *Road and Track* magazine comparison test in 1971, they said, "No car has ever come in more resoundingly last in a car-to-car evaluation."

The Super Beetle was a product of desperation, and so it was destined to fail. Volkswagen AG had been sitting pretty through the '50s and '60s and didn't notice little cars from Honda, Datsun and Toyota sneaking into Los Angeles ports and, more importantly, into people's garages. Quick, nimble and cheap, the Japanese onslaught of the compact car had begun, not to mention the domestics such as the Pinto and the Vega that slowly chipped away at Volkswagen's stronghold on the market.

The Standard Beetle was wearing out its welcome in America. People were moving on to other things. The egg car no longer enjoyed its share in the market as it did before. It needed to be replaced, but how? The Type III was unsuccessful and the Type 411 and 412 were considered a blight on VW AG's bottom line. The only thing VW had going for it was the Transporter line, but even its sales were fading fast.

Enter the Super Beetle with its MacPherson struts, more trunk space, 63mm wider front track, smaller turning radius (even though the wheel base was extended by 20mm) and double-jointed half-shafts. With all the improvements it was still stunted by the 1300cc engine re-tuned to produce 46hp, and even the 1600cc that provided 60hp wasn't much of a big change. These engines provided, at best, a 21-second quarter-mile at 64 mph, largely due to the new 2300-lb. weight of the car.

That's a pretty rocky first impression for the world, but the Super Beetle didn't get a fair chance from the start. The expectations were very high. And the Beetle, which had been around for more than 25 years at this point, was no longer considered very unique. Consumers were less interested in having something different, more interested in getting from one place to another as fast as possible with less fuel, which the Super Beetle could not do as well as its competitors. So the Super Beetle died five years later, only to resurface as a great used car, a perfect first-time vehicle for teenagers in search of wheels.

When people find out that I own a 1971 Super Beetle, they instantly think of two things: 1) That I got it because someone gave it to me (or I got a really good deal), or 2) It's for sale. Most are surprised that it was the first car I ever owned, and more people are surprised I still own it, and proudly at that.

I purchased the Super on April 8, 1989, four days before I turned 16, and I never even knew it was a Super. I didn't know what MacPherson struts were. I didn't know that the spare tire was not supposed to be horizontal and I had no idea how to decipher the VIN number. I drove it in ignorant bliss, like a lot of people with their first car. I didn't know much about Volkswagens back then (or cars in general), but I knew what I liked and what I didn't like.

I drove it until August 8, 1991, when I started college. After that, it sat safely in my folks' driveway, waiting until I found the time and the money to restore it.

That time came in June 2003 when, while at *VW Trends* magazine, I started Super Project '71 in response to the hundreds of technical questions we were getting monthly asking about Super Beetles. I carefully chronicled the project in a series of articles, which now appear here in this book. All of them were written by me except for the first chapter on the history of the Super Beetle written by Wayne Dean.

Volkswagen took a chance back in 1971 that the future of the company, at least for a while, would be saved by the Super Beetle. Well, as older models prove more and more difficult to find, we'll discover that it is the Super Beetle that holds the future of the VW hobby is in its hands.

Bulbous, awkward, bulky, yes, but it is still a Volkswagen.

—*Ryan Lee Price*

THE RISE AND FALL OF THE VOLKSWAGEN SUPER BEETLE
By Wayne Dean

Super: 'sü-p_r [adj] 1. excellent, extremely good, wonderful 2. extra good or large of its kind 3. above, beyond or over

In 1968, the Volkswagen Beetle was still selling over one million units per year in the USA. Heinz Nordhoff, the chairman of Volkswagen, was seriously ill, but even then knew that the future of Volkswagen could not rely on the mighty Beetle forever. The Type IIIs were not the saving grace that they were thought to be and something would have to be done soon. Sadly, Nordhoff passed away April 12, 1968, before the new Type IVs could come to market; another series that missed the mark for Volkswagen. Kurt Lotz took over the helm of Volkswagen, which at that time was like a ship without a rudder. There was a lot of uncertainty at Volkswagen as they watched the orders for the Beetle start to dwindle away.

U.S. automakers had learned a thing or two from Volkswagen. The new Chevrolet Vega and Ford Pinto gave U.S. car buyers the price, room and fuel economy that was needed at a time when gas prices were rising. The Japanese had also become major competition for Volkswagen. *Road and Track* magazine did a head-to-head comparison between the Beetle and the Toyota Corolla. Unfortunately for Volkswagen, the Corolla did it all better and for less money. This was embarrassing for Volkswagen, as the Beetle lost in every category except perhaps for its charming personality.

Of course, the Super Beetle's claim to fame, as shown on this 1971, is their increased trunk space. The strut towers allow the spare tire to lay down flat instead of upright as on previous models.

Until the new, water-cooled Rabbit was in production, VW decided to rework the Beetle and bring it up to the 1970's standards. The plan was to reduce the price of the standard Beetle and introduce a new updated model: a Super Beetle. The decision to produce the Super Beetle was going to be a costly one. Never before had Volkswagen spent so much time or money on the Beetle. The new suspension design required a new chassis and every panel in the front end had to be redesigned as well. This meant new, rounder fenders, a larger, wider hood, redesigned front apron, a new spare tire well, changes to other areas of the body and new inner fenders to accommodate the mounting of the new struts. This would be the most costly and extensive revision of the Beetle since it was first launched. If Heinz Nordhoff was around, it is said that the Super Beetle might never have been built.

The newly designed 1971 Beetle was to be called the 1302, or the

THE VOLKSWAGEN SUPER BEETLE HANDBOOK

The 1971 Super had a few modifications: a three inch longer body, twice the trunk space, a flow-through ventilation system with a two-speed blower and a 60hp engine.

Super Beetle. The 1302 designation was chosen due to the fact that car manufacturer Simca already had a model called the 1301. The Super Beetle was to be sold in North America with a carbureted 1600cc, 60hp engine and available in Europe with a 1300cc version as well. To define the 1600cc model from the 1300cc in Europe, an "S" was added, making the Beetle with the larger engine the 1302S. Planned for release in August 1970, the new brochure stated, "And now the new VW 1302S. The 1600cc Super Beetle. The most powerful, most exciting and most comfortable Beetle ever."

For the first time in the history of the Beetle, the spare tire was stored horizontally in a recessed wheel well under the cargo area in the front trunk floor. The jack was moved to under the rear seat and the air pressure powered windshield washer bottle was relocated to the right inner fender. These simple changes resulted in nine cubic feet of storage in the trunk, an increase of 86 percent. When this new space was added to the storage area behind the rear seat, the Beetle finally had the carrying capacity that people wanted. The new 1300cc and 1600cc engines featured dual port cylinder heads for better performance. And to help overcome the problem of keeping the number three cylinder cool, an external oil cooler was added and the tinwork was redesigned to allow more fresh air in. The rear deck lid was increased in size to accommodate the new larger engine and had two banks of five louvers cut into it to help keep the new power plant cool. Crescent-shaped air vents, trimmed with a silver metal edge were added behind the rear windows. These vents were part of the new flow-through ventilation system that was added to ensure adequate fresh air to the interior. Another nice touch was the addition of a passenger side vanity mirror to the sun visor. These small touches were part of what Volkswagen hoped the American public was looking for.

Part of Volkswagen's master plan was to bring the handling of the

For the first time ever, the world's most popular car came in two distincts versions, the standard and this 1971 Super Beetle.

THE RISE AND FALL OF THE VOLKSWAGEN SUPER BEETLE

1972 was a watershed for safety features on the Super: Dash features controled from the steering column, larger rear window and an energy-absorbing steering wheel.

Beetle up to North American standards with improved front suspension and updated rear trailing arms. MacPherson strut-coil springs were coupled with transverse control arms and a better turning radius was one of the results. The new suspension was lighter than the traditional torsion-beam design, the inner fenders were now heavier and the anti-roll bar was made much larger. The chassis frame head had to be substantially modified and made flatter to accommodate the new suspension changes. This suspension setup was similar to that used on the Type IV. In the rear, double-jointed half shafts were introduced. These were formerly only available along with the semi-automatic transmission. The 1302 now handled more like a Ford than a Beetle.

The 1302 sold surprisingly well despite Volkswagen enthusiasts giving it mixed reviews, with descriptions like "ugly," "swollen," "bulging" and "pregnant." With optional air conditioning for only $267 and a semiautomatic transmission for an additional $139, the Beetle was now offering creature comforts that would help to boost sales. Having Volkswagen's excellent reputation, dating back to the 1950s, didn't hurt either.

The 1970s were also the beginning of the special edition. With models such as the Sports Bug, Sun Bug, Love Bug, Fun Bug, Winter Bug, La Grande Bug, Champagne Bug and overseas, the Jeans, Big and City models. Another special edition for 1971 was the Jubilee Beetle celebrating over twenty million Volkswagen sales worldwide. This was one of the first special editions based on the 1300 standard and the 1302 Super Beetle. More than 35 different, special-edition Beetles were to follow, with even more being produced in Mexico today. 1971 also meant the end of the standard Beetle convertible. Until production ended in early 1980, all convertibles produced would be of the Super Beetle variety.

It was 1972, the second year of the Super Beetle, and it brought

One quick way to tell if it is a Super is by looking at the front apron. Super Beetles equipped with A/C featured these louvers to cool the condenser.

THE VOLKSWAGEN SUPER BEETLE HANDBOOK

with it some changes to the already new design. To increase visibility, the rear windshield was enlarged by 4cm (11 percent). This was to be the last time that the rear glass was enlarged and compared to the Split and the Oval window Beetles, the glass was now massive. The wiper switch was moved to the right side of the steering column for convenience and the vents in the rear engine lid were increased in number. Volkswagen's obsession with keeping the engine cool now required a massive 26 louvers. These were grouped in four unequal banks at the top of the rear lid. Continuing Volkswagen's strategy of offering people more luxury, there was now a shelf installed behind the rear seat. This cover could be hidden away or extended over the rear storage area in an effort to keep valuables away from prying eyes. A new, flat design, four-spoke, plastic steering wheel was added to the interior. This was to help prevent injury in case of an accident and came complete with a Wolfsburg emblem in the center. To help keep the flow-through ventilation draft free, a pair of vents were added to the dash, complete with tiny directional regulators. A diagnosis plug was installed in the engine. Customers were encouraged to bring their Beetles back to their local VW dealer for service. The dealer could then hook the Beetle up to a special, VW-only, computer analyzer and automatically check a number of areas on the car.

On February 17, 1972, the 15,007,034th Beetle was sold. Volkswagen had now claimed the

The interior of the 1973 Super Beetle, like all the Beetles that came before it, is Spartan but very functional.

A side view of a 1973 Super Beetle.

world production record for the most produced single make of car in history. It was a 1303 Beetle that took the honors, beating the 60-year-old record set by the Ford Model T. This was an important mark in automotive history and the car that hit the mark was donated by Volkswagen to the Smithsonian Museum in Washington, DC, for permanent display in the industrial section. On this special occasion, Volkswagen equipped six thousand Super Beetles with a blue, metallic paint scheme, special 10-spoke, pressed-steel wheels and gave it the title of Marathon Beetle. This European-only Beetle was the World Champion (it's called Der Weltmeister in Germany).

In the U.S., Volkswagen released 1,000 units called the Baja Champion SE. This model was to commemorate Baja off-road race successes from 1967–'71. The Baja title came from the desert races that were held each year in the Baja

THE RISE AND FALL OF THE VOLKSWAGEN SUPER BEETLE

A clean 1973 Super Beetle.

California region of Mexico. The fact that the Baja Champion was a 1302 was ironic because any Beetle that would have competed in this race would have used the tried and true torsion bar front end that had been in service on Volkswagens since 1935. For any would-be Volkswagen purchaser that wasn't lucky enough to get one of these limited production Baja's, you could buy the dealer-installed option package for $129.95. Included for that low price were side body decals with the Baja name, mag wheel-style hub caps, a special shifter, Bosch fog lights for the front bumper, walnut-colored trim for the dashboard, bumper guards and chrome taper exhaust tips.

Volkswagen was busy pushing the Super Beetle with a 14-page, full-color sales brochure with a cover that read, "The Super Beetle. The older it gets, the better it gets." The 1302 was available in seven colors, while the standard Beetle came in only four colors and had a four-page, black-and-white brochure.

In 1973, the cover of the eight-page color brochure for the Super Beetle read, "The '73 Beetle. All small cars are not created equal," a fact that Volkswagen had proven with years of constant improvements to the Beetle.

The changes that took place with the 1302 were just the beginning in the plans to modernize the Beetle. In 1973, Volkswagen took the next step with the introduction of a new-and-improved Super Beetle, called the 1303. VW purists could not believe their eyes when a redesigned, full-sized, padded dashboard replaced the traditional flat one that had been in the Beetle since 1958. This new dash was originally designed to house future air bags and to improve ventilation inside the car. By using an air vent channel that stretched from side to side near the front windshield, a greater volume of fresh-and-heated air could be delivered to the occupants. The single gauge still remained, but was now housed in a plastic binnacle in front of the driver. The other switches were moved downward, directly in line with the radio and right at the driver's fingertips. The glove box in the new 1303 was a good size but for some reason, was divided into smaller compartments. The lid would no longer open fully to double as a beverage tray. There was also a new fuse box located centrally for easy access in case of a failure.

In the rear of the Beetle were the largest taillights ever installed on a Volkswagen, and most likely any other car of that era. These soon earned the nickname "elephant's feet" in VW circles and were thought to be ugly, compared to the stylish "tombstone" taillights that preceded them. Everything that was changed on the Beetle was done so for function and style and the new windshield was no exception. To comply with proposed U.S. safety regulations regarding the distance between passengers and the front windshield, Volkswagen introduced a sharply curved front glass to the 1303. This panoramic glass gave a remarkable 42-percent increase in visibility and improved the aero dynamics of the car as well. It also caused a change in the shape of the front hood and the roofline. The new shortened hood gave the car a pregnant look and lost its VW circle logo in the transformation.

One of the nicest special editions for 1973 was the beautiful Sports Beetle. For an additional $250.00 this 1303 Super Beetle came with distinctive red and black stripes that encircled the car. The tapered tailpipe tips, trim, door handles, wipers and bumpers were all given

THE VOLKSWAGEN SUPER BEETLE HANDBOOK

a matte black finish, and 5.5-in silver Lemmertz GT wheels with radial tires were installed. The Sport Bug's sales brochure featured black headlamp rings although some of the cars were fitted with the standard chrome ones. The interior came with sports bucket seats, a leather sports steering wheel and gearshift knob.

In 1974, the last Beetle rolled off the production line in Wolfsburg to make room for the new Golf model. As usual, Volkswagen brought out a few more changes for the Super Beetle. U.S. regulations now required that every car be able to withstand a 5 mph front and a 2.5 mph rear impact and sustain no damage. To comply with this new hurtle Volkswagen added what were called self-restoring energy absorbing bumpers. Impact-absorbing shocks were added to the now thicker steel front and rear bumpers to accomplish this task. To ensure the front occupants were wearing their seat belts an ignition interlock was installed so that the car could not be started unless the seat belts were fastened. Fortunately this system was quite easy to disable by merely unplugging the sensors under both front seats. To improve handling under hard braking, the Super Beetle now had a negative kingpin offset. Also in 1974 the old style generator was finally replaced with an alternator, and a new type of alloy was used to improve the cylinder head life.

The 16-page 1974 sales brochure featured both standard and Super Beetles. On the cover was a picture of a red 1303 Beetle floating in the ocean with the caption, "The VW

The cutaway drawing of the 1974 Super shows the energy-absorbing bumpers, four-wheel independent suspension, air-cooled engine and one of the sensors that is part of the computer analysis system, the probe leading into the battery for testing electrolyte levels.

Beetles. Built better than ever." This would be another year for the special edition with the introduction of the famous Sun Bug. Available in standard, 1303 sedan and convertible models, the Sun Bug was painted a beautiful Hellas Metallic and came fully loaded. Stamped silver sports wheels, Kamei tunnel console, a sunroof with wind deflector (Sedan only) and wood-finish dash panels were just some of the special features. The seats had special Nut Brown upholstery, as did the matching door, side panels and loop pile carpet. There was a Sun Bug logo on the gearshift knob and

This 1974 Super is just one year after the major rear-end changes, such as additional louvers, energy-impact bumpers and larger taillights. Part of a protection program called "VW Owner's Security Blanket," the 1974 Supers were covered by a warranty that included everything but fluids, filters and tires for 12 months or 20,000 miles.

THE RISE AND FALL OF THE VOLKSWAGEN SUPER BEETLE

one installed on the engine cover by the dealer. The sales brochure for the Sun Bug even had the words to a song written especially for the car entitled "Let a little sunshine into your life" by Keith Konnes. The lyrics for the song went like this:

Fresh air and scenery. It's all there and it's all free.
Open your eyes, it's there to see.
Let a little sunshine into your life!
Keep your disposition from draggin' in a golden Sun Bug from Volkswagen.
Even make a short trip will set your heart a waggin'.
Let a little sunshine into your life.
Sun Bug Beetle or Super Beetle opens up a window to the sky.
Sun Bug convertible goes the whole route so you can wave at the world as you go by.
Great gas economy, easy on the world's ecology. That's the Sun Bug philosophy.
Smile your way through problems and strife and let a little sunshine into your life.

Improvements were in order once again for the Super Beetle in 1975. The worm and roller steering box was replaced by modern rack and pinion steering and improvements in the rear-end geometry were made. The engine case would now be made from a better alloy classified as AS21 and the twin tail pipes that had been on the Beetle since 1956 had now become only one.

Again U.S. regulations forced Volkswagen and other car manufacturers to clean up their act pollution wise. Unleaded gas

Unlike regular VW Convertibles of the day, this special edition comes in ivory paint, a white interior and an off-white top. A four-spoke steering wheel, wood-grain dash and sport rims are some of the special features.

would be the new diet for the Super Beetle as computerized Bosch L-Jetronic fuel injection was added. This boosted the Beetles fuel economy from 25 to 33 mpg. A silver "Fuel Injection" logo was added to the rear engine lid where the Volkswagen script had been for years. Beetles produced for California sported a catalytic converter under the now bulging rear apron, and pretty soon all the States would require one.

Sales dropped from 791,023 in 1974 to only 441,116 in 1975. Although this may still seem like a huge number of cars compared to today's sales figures, Volkswagen

Built on a chassis produced by Karmann, this 1976 Vert's fuel injection engine delivered 34mpg highway and 22mpg city.

The rear shot of the popular La Grande Bug in 1975. It featured electronic fuel injection for increased horsepower and fuel economy.

had been used to selling over 900,000 Beetles annually since 1964 in the USA alone. In February 1975 Tony Schmucker took over as chairman of Volkswagen. He decided to end production of the 1303 sedan. Only the Super Beetle convertible and a standard Beetle sedan were to continue.

In September 1975 the new Golf GTi debuted at the Frankfurt Motor show. Its powerful water-cooled 1600cc engine put out 110bhp. That was more than double that of the Super Beetle. It seemed the writing was on the wall for the European production of the Beetle, as more plants were needed to produce the newer models.

The year 1975 produced a number of special edition Beetles as well, one of which was called the La Grande Bug. This 1303 Beetle was basically a reworked Sun Bug, minus the logos and trim. The sale brochure for this one proudly exclaimed, "You don't drive in it, you arrive in it" with a picture of a chauffeur in uniform in front of an elegant mansion.

In 1976, the 1303 sedan had faded into history but the 1303 Super Beetle convertible remained. There were very few convertibles being produced by U.S. automakers and the Super Beetle was one of the only choices for those who wanted the open-air experience. Worldwide orders started to increase, so the order was sent to Karmann in Osnabrück to increase production of the convertible from 33 to 50 units per day. The increase in sales sparked the release of one of the most popular special editions, the Triple White convertible. White was Germanys national racing color so this 1303 came with Alpine white paint, Opal (white) upholstery and a white convertible top.

U.S. sales for the 1303 convertible had been increasing by over 5,000 units per year since 1974. And in 1977 The Champagne Edition was Volkswagens latest special offering. Like the Triple White the car was Alpine white with Opal upholstery, but this time it came with a light ivory (or light sand) colored top. There was a gold stripe that was applied just above the running boards on each side of the body, a rosewood dash insert, sports wheels and even whitewall tires. The 1303 Super convertible was one of the nicest equipped Beetles ever. Volkswagen made it even better with the addition of a rear-window defroster and adjustable front headrests.

On January 19, 1978, European production ended for the standard Beetle sedan at the Emden plant, but the convertible was to continue at the Karmann factory for a couple of years more. The 1978 sales brochure read, "Once again, Volkswagen promises you the sun, the moon and the stars." However, articles in magazines had people speculating that 1978 would be the last year for the convertible. The forced an increase in production to keep up with the demand for the soft top. Colors available were Chrome yellow, Mars red, Barrier blue and of course Alpine white. The Champagne edition was back as the Champagne II and for the first time ever it included a Blaupunkt radio, quartz clock and a burled elm wood appliqué on the dash.

Volkswagen surprised everyone in 1979 when the 1303 Super convertible returned for yet another year. The new federal safety and pollution guidelines had been delayed and that allowed the

THE RISE AND FALL OF THE VOLKSWAGEN SUPER BEETLE

The Super was available in many different guises, including this "La Grande Bug" with metallic paint, special upholstery and special wheels.

Beetle to pay North America its final visit. All of the options that were on the Champagne edition Beetle were now standard equipment. The 1303 was now the only four-seat convertible for sale in the U.S., and at $6,495.00 was still a good deal. The last special edition Beetle produced was called the Epilog convertible or simply the "Triple Black." The car featured black paint with a matching interior and top. The Epilog was produced to show the bond between the first Kdf–Wagen convertibles assembled 40 years prior to this date that were also painted black. Included for the $200 extra, that the Triple Black option cost, was an am/fm radio that was added to the other standard equipment.

The news started to spread that this was the last year of production for the Beetle convertible. This created a huge backlog of orders at the Karmann plant. Production that was supposed to end July 31, 1979, was kept up until January 1980 to fill the thousands of orders that poured in from around the world. But on that fateful day January 10, 1980, the last Super Beetle, a triple-white convertible, rolled off the production line and into history forever. This car can be seen today on display in the Karmann museum in Onasbrück, Germany.

The very last one-page brochure for the 1303 Super Beetle simply stated,

"After 29 years, millions of Beetles, and countless improvements, the 1979 Convertible is still a very sensible way to flip your lid."

2 INITIAL INSPECTION AND EVALUATION

The Super Beetle is a funny car. It is inflated, rounded, bulbous and bloated, but lots of people love it. It looks solid, squat, thick and sturdy. Its name is an adjective, meaning better than everything else, and implies reliability. It's longer, faster, wider, taller, roomier and heavier. It's a Super Beetle.

The Super Beetle was introduced in 1971, and quite a few of them were sold. Of that number, only 331,191 Beetles of all types made their way to the U.S., and Canada, including this one. It was built in Germany in late September, 1970, and originally sold at Valley Autohouse, Ltd., in Abbotsford, B.C., Canada, which is still a VW dealer today. It made its way south in the early '80s and was purchased on April 8, 1989 for $1800. Of course, it was in much better shape even in 1989 than pictured here. Time, weather, sun, heat and moisture has taken quite a toll on its parts, leaving it in this less-than-perfect condition. Nonetheless, it will provide a great platform for any project.

If you've got a Super Beetle laying around and it looks at all like this one, chances are you want to return it to its former glory. Well, you're in luck, because in the pages of this book, we plan to put this Beetle back into shape. So get out your tools and be prepared to follow along.

Fifteen years baking and soaking in California weather has taken its toll. We have a big job ahead, but it will also be an adventure.

Throughout this book, we will try to explain the ins and outs of the various steps along the way. Though they may differ from your idea and order of how to do things, we will cover all aspects of the buildup in a useful and informative way. The chapters to follow include a complete rebuild of the suspension systems, both front and rear, a complete rebuild of the brake system, from the pedal to the drums, a fully refurbished interior, paint and body, electrical, engine and transmission and an overview of all of the little things that make a Super Beetle so super. For example, that crazy emissions control system canister and collector found in the Super's fuel system, the MacPherson struts and rear semi-trailing arm suspension that had been offered on the '67-and-later automatic. Of course, along the way in our project, we'll discover problems and obstacles that need to be dealt with before we can reach our final goal: to drive off into the sunset with a

INITIAL INSPECTION AND EVALUATION

A nice front-three-quarter shot shows a fairly solid Volkswagen. The roof rack will have to go, as it belongs on a '67, and won't go with the overall theme of restoring this car back to stock form.

The rear of the car is in good shape but, like the rest of it, it needs attention. We were lucky to have most of the accessories and stock equipment still bolted to the car.

The engine was removed several years ago and all that's left is an old transmission, air hoses and various wires that will all need to be sorted through and cleaned.

Our first clues to this car's past are these matching body panel weld marks on their side of the engine compartment. With "Mexico" stamped on the rear apron and fenders, it is clear that the car was hit hard enough in the rear to necessitate a replacement. It also appears as though the work was done by a VW dealership.

We hadn't seen Gabriel's Roadstar shocks in a number of years (as they discontinued offering this model in the '80s), and the axle boot hasn't ever been worked on by a VW dealer, because, after 1972, the driveshaft featured a ridge that better secured the small end of the boot as well as a clampless rolled seal over the end cap. The use of hose and pinch clamps were discontinued and often upgraded on later models.

beautifully restored/rebuilt 1971 Super Beetle.

OVERALL INSPECTION

First, let's take a walk around the car and look at some of the immediate things we'll have to attend to as this project develops. After being left outside for most of its life, the body has begun to rust, the rubber window moldings have started to crack and most everything plastic is brittle, shattered and/or deteriorated. In the few years I drove it as a daily driver, it never had a flat tire and never needed one replaced, but in the 10 years it sat under the sun and rain, two sets of tires rotted out from under it.

Underneath the cracked, oxidized paint, overall, this appears to be a very straight and solid car, with no major scars to worry about—only a couple of minor ones we'll point out in due time—no previous bodywork that needs to be fixed and no horrific "restorations" to undo and redo.

Starting in the front, the first thing that catches your eye is the

THE VOLKSWAGEN SUPER BEETLE HANDBOOK

The once chrome bumpers have rusted so much that they cannot be merely cleaned. Either they will have to be rechromed or replaced.

The running boards will also need replacing. The corrosion even claimed two connection bolts. As well, one of the bolts between the front fender and the board had to be cut off. I see new ones in my future.

Just some of the dings and dents that will have to be removed. Not shown is a slight wave on the hood and a fist-size dent near the right-side cowl. The hallmark of a teenage driver!

Underneath the driver's side fender we find a host of things to add to our to-do list. A spring/strut job is first. Then we'll find out why there is a couple of inches of play in the steering wheel. The tie rod ends will all see the trash can and all bushings will be replaced.

Though it's still round and still holds a good bead with the tire, this wheel will get sandblasted and repainted.

A big place for rust is in the door hinges. Care must be taken when removing them at this point until we can find out how serious the damage is. Don't knock out the pins with a hammer, whatever you do.

surface rust along most of the hood's ridges and on the apron, especially around the large dent I added as the result of an off-road excursion I took one day. While we're down there, you may have noticed the lack of the 39 louvers that are normally associated with a Super Beetle's front apron. In most warm-climate countries these were added to ventilate a condenser mounted behind the apron, part of an optional air-conditioning unit. Generally, it wasn't part of Canada's line, and was offered as factory option M559 instead. To give further credit to this is an Eberspacher gas-fired heater (311-261-103—factory option M60) stuck under the hood. However, on this particular car, it never worked. Plus, the idea of keeping a flame that close to 42 liters (11.09 gallons) of fuel was always a little unnerving, so it probably won't be returned to the car when the time comes. But like any project, things change along the way. For example, this project was originally slated as a big-motor German Looker and now we're opting for a factory stocker.

The front bumper's rusted, but straight, and the rubber impact strip (M162) is cracked in several places (one from running into a fence—oops). The bumpers will have to be either rechromed or replaced altogether, as a spot of steel wool has proven useless on several occasions. The tires are shot...again. The wheels are rusted, flaking and worn in places, and the running boards' rubber covers have split. The door hinges have rusted,

INITIAL INSPECTION AND EVALUATION

The dashpad is cracked and brittle. The metal around the radio hole is scarred and this dirty worn steering wheel must go.

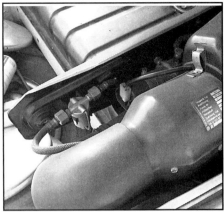

One gem in this car is the Eberspacher gas-fired heater (311-261-103—factory option M60). Though it never worked, it is a sought-after item by purists. I think we'll let some other lucky soul try to figure out how to get this thing working again. Of course, removing it means filling three holes in the body.

Just north of the gas heater is the brain of the whole car; behind the speedo pod is this collection of wires.

Underneath the back seat was this conglomo of useless wires, most of which led to two aftermarket horns that were on the car.

And to the left is the battery tray. Since this is a black and white, I'll describe the color: Rust. The positive cable corroded to the battery terminal and had to be cut off, while the metal lip that secures the base of the battery is rusted as well.

the window scrapers are hard as rocks, but the sash works well, rolling the window up and down as it always had. The roof is straight and dent free, though some rust buildup has collected in the sills.

The bottom of every body panel has some degree of rust—the fenders, the doors, the aprons and the quarter panels. The right rear fender has a large dent (that's rusted over) because a kid ran into it on his bicycle, and it'll have to be fixed.

The rear bumper is the same as the front, but was replaced once before in its life, because it is missing the rubber impact strip (from the luxury package started in 1970) and, instead, has the black strip found on most all Super Beetle bumpers. There is a "Mexico" stamping on the rear apron and some weld marks on the inside, suggesting that the apron has been replaced by an aftermarket unit, and each rear fender has "Mexico" stampings as well. On the surface, the rear end bodywork was done good and straight, and I've seen worse. Since the engine's gone, we get a nice view of the compartment cluttered with spider webs, leaves and an assortment of random wires, hoses and trash that will have to be gutted and sorted. To be on the safe side, the transmission will probably be replaced.

Mechanically, most of the brakes' parts will need to be replaced, as all of the lines are brittle and one

Speaking of rust, one of our fears is connected to this image of water damage on the headliner underneath the rear window. Without pulling up the carpet, it sounded a little crunchy when we pressed on it.

A closer look at the knobs and controls on the dash show that they are, for the most part, in good working order. If yours are also, be glad, because they don't make new ones anymore.

See my welding handiwork on the broken clutch pedal? I did this when I was 17 years old, and I still have the burn scar on my hand to prove it. As proud as I am of my work, I still want to get a new pedal cluster for safety's sake.

This is a closer look at the right wheel hub, and somebody was good enough to scribe an "R" on there, for "rust" no doubt.

Most all of the bushings underneath the front end are in one degree of disrepair or another. The track control arm, the steering knuckle, the ball joint and the stabilizer bar need attention, as well as do their corresponding bushings.

Through the cob webs is the drop arm which is part of the steering system. The top of this photo points to the rear of the car. The drop arm is attached to the base of the steering gearbox at the top, while connected to the steering damper at the bottom. The steering damper prevents shock from the road to be transferred through the steering wheel to the driver.

metal line is even crimped somehow. The drums are rusted and flaking, the shoes always gave some trouble and the cylinders have long since dried up. Pressing on the pedal means nothing to this car anymore. The e-brake line works to some extent, as long as you don't give the car a good shove. The struts and shocks, ball joints, CV boots, steering arms, steering box, knuckles and tie rods will need a complete going through. A light film of rust covers everything underneath the car, but only breaks completely through under the front right fender (nearest the curb, hence, nearest the water-filled gutter).

The interior is a mess and always was, even when this car was running. With the exception of the front and back seat frames, everything will have to be replaced. The carpet (house variety...it was a thick tan shag if I remember right) is long gone. The seat covers have ripped through, showing the original covers underneath. The dashboard pad has cracked in many spots, and when I went to adjust the rear-view mirror, it simply crumbled in my hand.

In the back, the rear window has leaked over the years and stained the upper portion of the headliner. Hopefully it hasn't caused any damage to the package tray (though I did hear something crunch under the carpet). Underneath the battery, which is long gone, it looks like Mars, red, rusted and rocky with oxidized bits of flaking metal. There are wires

INITIAL INSPECTION AND EVALUATION

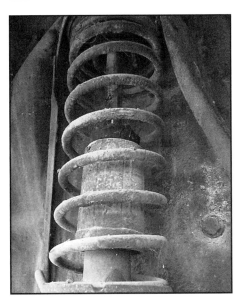

The coil spring on the strut looks fine...but we won't know for sure until we can take it apart.

Here's what years of water, salt and the elements can do to metal. These panels will have to be fixed.

The whole double-jointed rear axle unit, from the transmission on the left to the wheel on the right.

A closer look at the end cap, CV joint and screws that hold it all together. It looks similar to something found on the *Titanic* at the bottom of the Atlantic. But it is all easily replaceable.

This canister, located under the right rear fender is part of the evaporative emission control system. Supposedly, this canister is filled with charcoal, which absorbs vapors from the fuel tank and pushes them through the carburetor. It doesn't work, and I know so because the line that connects it to the expansion chamber (under the cowl) is disconnected.

everywhere, like an electrician went nuts. I had some fun running speakers all throughout this car, as well as a couple of horn buttons under the dash (one for the regular horn and another for an oogah horn). The steering wheel is a Grant GT knockoff that will be replaced, and the pedal cluster will need some attention, especially since I had to weld the clutch pedal back on after it broke off one day, which made driving it quite difficult. I'm not much of a welder, so it would make me feel better not to have to do it again. And, as a novice, stereo-crazy kid, I just had to have a fancy new cassette player installed, even if it meant making 14 separate cuts in the dash to make it fit. So I have some welding to do whether I like it or not.

GETTING STARTED

This is what we have to deal with so let's get started. But where do we start? Good question, as there is a hundred different ways to begin, and whether you begin by gutting the car or pulling the body from the pan, it makes no difference as long as you plan it out and stick to the plan.

Wheels and Tires

We're starting with tires and wheels. I know, it sounds rather backward, but since I had to chain the Super to the truck and drag it out of the garage on flat tires and rusted, stuck drums, I need to get it physically rolling again. Also considering that we will need to take it to several different places during the build-up process, we wanted to support it with some good road wheels and tires during all of the towing. Be safe once instead of sorry twice (you know what I mean).

New wheels for a four-lug Beetle can be found most anywhere, but we want to retain as much as we could of the original car, so we decided to sandblast and powdercoat these. There are several other options, such as a wire brush, some sandpaper, solvent and a couple of cans of L91 chrome silver, but we chose powdercoating instead. It's cleaner, quicker, the surface is protected from the elements for a longer time, and you can get any color you want. If you have to think economically, try the DIY kits from Eastwood, as powdercoating is sometimes an unnecessary luxury.

If it is a must for you, odds are good you've got your favorite sandblaster and powder coater's phone number on your speed dial, so let him know you'll be by with a few old wheels. Mine is Gerhard Schapp of G&M Schapp Powdercoating, who is in partnership with Ron Rose at R&R Sandblasting, conveniently located next door to each other. Schapp is an old-school German who's been around VWs his whole life, while Ron knows exactly what we're looking for in a set of refurbished wheels. Out of one shop, into the other and back home on the car in only three days; it was a remarkable transformation from rusty, dirty and spotty to clean, shiny and new.

It's hard to tell if the starter is in working order or not. Odds are good that it is serviceable, but we won't be able to tell unit the unit is removed or an engine is added.

The only thing we actually did at this stage was to powdercoat the wheels and mount new tires. Since we are planning on a stock Super restoration, we chose L91 Chrome wrapped with 165/15 1/2-inch whitewall BFGoodrich tires.

Keeping as close to stock as possible, we went with 165/15 1/2-inch whitewall BFGoodrich tires from Coker Tires, which was a dealer-installed option at the time. Coker Tires have always been a good source for period-correct rubber for almost any type of vehicle, so there is never a question where to go.

CONCLUSION

Now that the car rolls, pushing it in and out of the garage on fresh tires will make working on it a lot easier. In the next chapter, we'll go though the highlights of tearing out the interior, the gas tank, wiring and everything that won't eventually get painted. As well, we'll discuss some problem areas we encountered and how we plan to overco me them—more importantly, how you can too.

INTERIOR/EXTERIOR DISASSEMBLY 3

In the last chapter, we got a first-hand look at the overall condition of our '71 Super, which gave us a good starting point. Now we need to completely disassemble the car. That would seem to be a simple task; just get a blowtorch, a lot of wrenches and start tearing it apart. But, like everything else, disassembly should be done with planning. There's a lot you can learn as you strip the car apart, and it will reveal unexpected repairs and damage you didn't see on the initial inspection. And, you need to record things as they come off so you know how to assemble them later on, which might be months down the road. But that's not to say it can't be fun either.

In this chapter of the book, we will begin by tearing apart the interior, removing everything from the inside of the car that isn't going to get painted. As well, we will pull out the gas-powered optional heater in the trunk, the fuel tank, air-ventilation system and the wiring. Since the engine's gone, we don't have to worry about much at that end, but we will need to remove the exterior badges, taillights, bumpers and rubber seals, not to mention the windows and the headliner. All in all, since there's a lot of work to be done, plan on getting plenty dirty in the process.

We started by rolling the car out into the driveway for ventilation, some sunshine and ample light to see by. You can see that our Craftsman tool chest is handy, as is a shop vac, sander and various tools any garage should have. It's time for the fun.

GETTING STARTED

It is advisable to save everything. Get a bunch of boxes, crates or shelves in your garage and start labeling parts. We used a few dozen freezer-style baggies for little parts and grouped like parts together so we knew what we had and where it came from. There were certain things we knew we would need to replace, such as the rearview mirror and the dash pad because they were both virtually destroyed upon removal, but they still earned a place on the shelf. It is better to have the old part handy while looking for a replacement (if necessary) rather than not have anything to go by during your search. As well, there are some things on a Super Beetle, for example, the steering column universal joint boots, that are no longer available, and cleaning up dirty, slightly damaged ones is better than not having any at all.

Another thing: Take ample notes and pictures. It can get complicated sometimes, especially when you're putting everything back and you forgot where one of the half-dozen air vent hoses goes. Draw diagrams if you'd like.

As I said, you can discover damage that you didn't see at first, which is what happened with us when we removed the rear seat.

THE VOLKSWAGEN SUPER BEETLE HANDBOOK

This is where we ran into our first major problem—major rust damage. The more carpet we uncovered, the more rust we found. So much so, if it had an engine under there, we could have almost adjusted the carb through the holes. In England, they call this problem "teabag," and you will see why. What rusted parts of the package tray that didn't pull up with the carpet were easily pushed through with slight finger pressure.

Removing most of the car's features is a straightforward ordeal and can be done with the simplest of tools, as illustrated in the pictures.

CONCLUSION

With everything torn out of the Super, we were left with boxes of dirty parts and an empty shell, aside from the steering wheel, pedals and door and vent windows which remained in the car. As well, we decided to leave the door handles and striker plates so we could still secure the door and not have to worry about tying them down each time. We'll show you the R&R of these in Chapter 13 before we begin to paint. The next day, we towed the Super Project 71 down to John Chabot at Topline, the best Super-Beetle-only shop in Southern California, to begin the restoration.

We'll pick it up there in the next chapter, as John takes a look at what we have to work with, recommends a few modifications and tests the reliability of our suspension and steering systems.

Window removal is easy if you take your time and be gentle, after all, it is just glass. Peeling up the aluminum trim exposes a nice channel to run a utility knife through, cutting away enough rubber to free the window. Careful, the front window is always brittle and easily breakable.

Removing the door panels requires that first, the window crank and door latch both be removed. Pry the plastic trim off of the window crank and the plastic finger plate out of the door latch. Since these screws are both protected by Loctite®, you'll hear a small snap before they loosen.

INTERIOR/EXTERIOR DISASSEMBLY

We found this scrap of material hiding behind the door panel. It was used between the door latch tether and the door frame to minimize rattling. Perhaps it is a piece of the car's history.

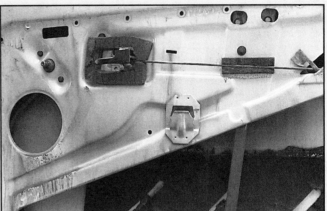

Once the handle and crank are removed, use something flat (a plastic kitchen spatula for example) to pop the 20-odd tabs that hold the door panel in place. On 1973 and later, the armrest is bolted to the door frame and has to be removed, but before this ('72 and earlier) it simply lifts up. The rear quarter panels come out the same way, but then make sure to remove the flip-out ashtray (if yours is equipped with one).

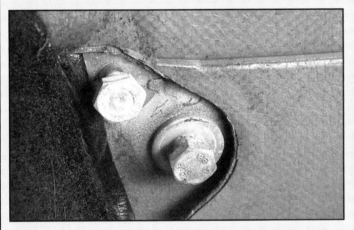

These are the seatbelt attachment and the rear seatback swivel point bolts. They both are 17mm and, once removed, free the seatbelts and the rear seatback. Don't lose the seatbelt bolts, as replacements are difficult to find.

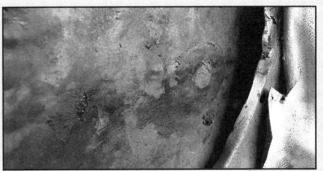

Behind the driver's side rear quarter panel we discovered our second problem. Unbeknownst to me, the car was apparently struck hard in the rear quarter panel during prior ownership. This makes for accident number two, and though the outside looks smooth (I hope there's no bondo), the inside is as wavy as can be. This also explains why the driver's door showed signs of a repaint (paint on top of the plastic). Add this to our "fix-it" list.

Two Phillips head screws need to be undone to free the foot well vents and subsequently the threshold carpet.

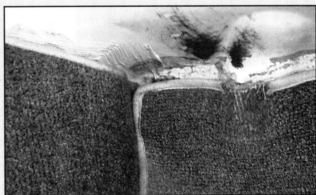

A prior owner used an excessive amount of glue to hold the carpet kit into place, as can be seen here. You haven't experienced tedious, hellish work until you've contorted your body under the dash of a VW and scraped at 20-year-old fabric glue.

The threshold pieces of the carpet come up easily by lifting them from a small groove that grips them from under the rubber strip.

With the rear seat removed, we got a good look at the carpet kit that was still in the back. Note it is a different color than the front. Attention to detail was obviously not the previous owner's forte.

And under the carpet was this scary mess.

The sound-deadening material on the vertical portion of the rear package tray was difficult to remove, but sandpaper and a wire brush helped. Remember to cover any exposed bare metal with a primer before leaving them for too long.

We don't want to scare you, but this is what we found underneath the rear carpet. This is going to require major sheet metal restoration. Hopefully, yours isn't as bad.

INTERIOR/EXTERIOR DISASSEMBLY

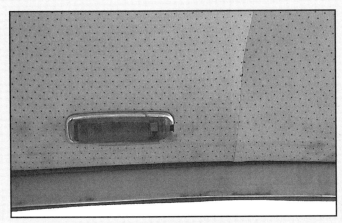

The headliner is next, but first you have to pull down the dome light. It is secured with a pressure clip that was easily removed. Of course, ours broke the moment we touched it, but in the bag it goes. It's a good idea to leave the wires alone so you can have a leader to run the new harness through the A-pillar.

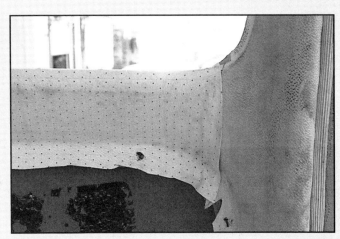

The perforated headliner surrounding the window simply peels off, while the leatherette headliner piece that covers the B-pillars needs to be carefully removed from the serrated spikes that clamp it down. Use a putty knife or the spatula to bend the serrations outward. Be careful not to break them or cut yourself, as they are sharp.

This is the front seatbelt attachment point. If you've got your camera out, it is a good idea to knock off a shot of this area. After we redo the headliner in Chapter 19, we will probably cover over this hole and lose it. When we put in the seatbelts (it's the law, you know), we'll need some help finding it again.

The 1971 Super Beetle was the first car that received a fresh-air ventilation system with a two-speed fan, shown here. Remove the fresh-air hoses that lead to the two ducts into the cabin. Three Phillip's head screws and a hex nut at the bottom free the box, but you still have to undo the wiring and the two Phillip's head screws that holds the control bracket to the back of the dash. Don't forget to remove the knobs, of course.

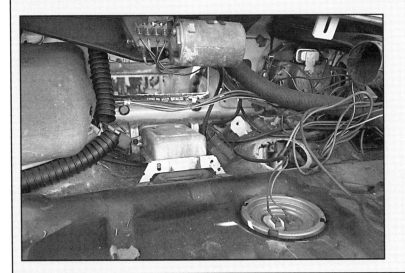

Without the cumbersome fresh-air box, we get a wide view of the mess behind the dash. Hoses, wires and stuff clutter the scene. At the top is the windshield wiper motor. Left is the glovebox and fuse box

THE VOLKSWAGEN SUPER BEETLE HANDBOOK

This confusing mass of wires was easy to remove, because most of them were speaker wires and those stupid horns. Some important components to look out for are the three relays attached above the fuse box. From left to right they are: Low beam relay, turn signal/emergency flashers relay and door buzzer relay. And remember to leave alone the main harness as it passes from the engine compartment to the main cabin on the driver's side to the left of the rear seats. It is an important connection that's nearly impossible to redo.

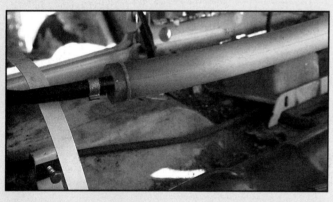

The working end of the Super's emission control system is this plastic "expansion chamber" located under the cowl. Fuel flows up into this tube when it expands from the tank. It then flows back into the tank when there is room. As well, a line forces expelled fumes to the rear of the car and into a charcoal canister under the right rear fender. This whole thing is supposed to stop fuel vapor and fumes from escaping into the atmosphere. To be correct, we'll have to get this working again.

The '71 Beetles have a new fuel tank, with a lot more hose than in previous models. When you remove these hoses, dump the clamps and get new ones. And make sure you mark where each one goes, otherwise you'll be stuck.

With all the wires gone, the trunk area is starting to clean up. Wow, look at that old Clarion speaker! That's gotta go.

From the rear, the interior is pretty well gutted. The next step is to clean off all of this glue, get rid of the stuck headliner sisal and vacuum everything spick and span. Let's tear out the dingy dash pad.

INTERIOR/EXTERIOR DISASSEMBLY

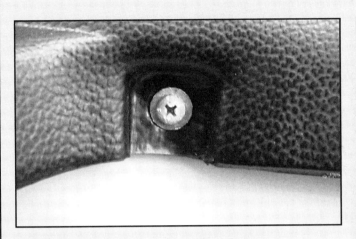

Along the bottom of the pad, as it wraps underneath the dash are six screws similar to this one that have to be removed.

Notice only half the dash came off. It's glued, yes, but it should have pulled up easier than this. We must have missed something.

And yes we did miss something. Behind the dash are two 12mm nuts like this one, one on either side of the trunk. The threaded part is attached to a metal strip that lines the inside of the dash pad.

After carefully removing all of the surface rust, debris and stuck tar paper and sound-deadening material, we painted the rear area with two coats of primer, one a deep etching sealer and the other a rust preventative primer (both from our local home supply store). There's still work to be done back there (as you'll see in Chapter 7), as this is only a temporary fix.

While we were in the painting mood, we sealed the floor pans and battery tray with two coats of POR-15. If you get any on you (or anything else), you'll have to live with it until it wears off, so we recommend latex gloves.

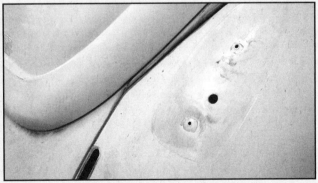

There's nothing unusual about anything on the exterior. The body moldings pop off with the spatula, the rubber grommets on the rear badge unscrew easily, as do the screws for the tail light housings.

23

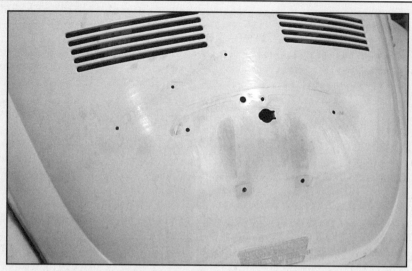

You may notice that this Super has the "old style" Super deck lid. Later in the production run, they changed the line to include 26 louvers (instead of the 10 you see here) and a fan-driven cooling system. The fan leaves a rectangle indentation right behind the license plate and helps ventilate the engine compartment.

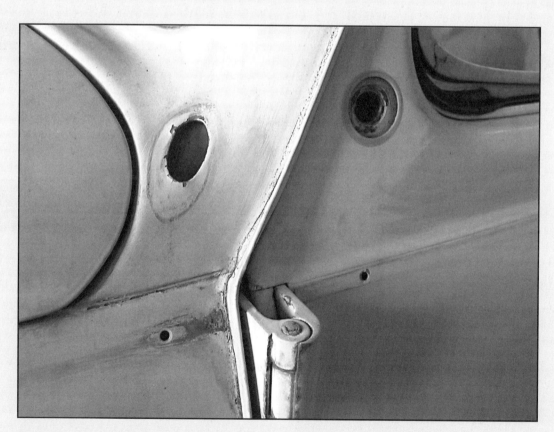

The mirror simply unscrews from the body. The antenna comes off by unscrewing a small threaded washer on the outside but is removed by pulling the aerial in through the trunk.

FRONT SUSPENSION DISASSEMBLY

The idea behind the work done in the last chapter was that the car was being ultimately prepped for paint, and the initial tear-down was taking shape. We experienced a few setbacks, namely a lot of rust, but it went smoothly. Jon Chabot at Topline in Anaheim, California, will handle the next few phases. The plan is to remove the front suspension and steering mechanism, check them for wear and clean them up.

The front end of a Super Beetle is like no other Beetle prior to 1971, as Volkswagen finally ditched Porsche's patented torsion bars and entered the modern era of car building. The new suspension allowed for more total trunk space (14.1 cubic feet from 8.9 cubic feet), a wider wheel base (95.3 inches from 94.5), a wider front track (54.3 inches from 51.6 inches) and a narrower wall-to-wall turning radius of 29.5 feet instead of the 36 feet of its brother Beetle

If there was one thing that sets apart a Super Beetle from any other Beetle, it would be the Mac-Pherson strut suspension. Though synonymous with the Super Beetle, the strut suspension is actually a system modified from the 411 to fit a Type I chassis. The coil springs (Part No. 113411105) surrounded a double-acting shock absorber and came in three different weights: one red paint stripe signifies pressure between 500–515 lbs., two red stripes, 516–530 lbs and

The first order of business, once the Super has arrived at Topline, is to jack up the front end and remove the tires/wheels and fenders to gain access to the suspension and steering.

three red stripes marked a pressure between 531–545 lbs. Odds are good that your Super's springs have two red strips underneath all of that dirt and undercoating. Check in the neighborhood of the third or fourth coil down. Both springs must match, so make sure you've got a pair before returning them to the car (though they're mostly indestructible, might as well get new ones, you're going to all of this trouble to remove them).

The springs were held by two plates at the top and the bottom, and a special spring compression tool is required to change out the springs. Important Safety Tip: Don't ever unbolt the nut at the top of the piston rod (under the ball thrust bearing dust cap). The pressure of the compressed spring can inflict serious injury to not only yourself, but to anything that happens to get in the way. If you don't have the proper tools, take the strut to someone who does.

The suspension strut fits under the wheel well by three 14mm bolts, and at the bottom to the steering knuckle via three 15mm bolts. The steering knuckle attaches to the tie rods and the ball joint, which is supported by the control arm. Sounds kind of confusing at first, but once you get down there and see it all, it's pretty clear. Though the mounting point of the suspension strut is different for the 1974 model, the principle is the same. Attached to the front frame head via two brackets (133-411-333) and rubber (or urethane nowadays) bushings (113-411-

313A) and to the control arms via two rubber bushings (113-411-315) and mounting plates (113-411-319) and nuts to the control arms is the stabilizer bar, which is unique to Super Beetles. This bar provides anti-roll properties but also helps keep the suspension struts in line.

The steering system for a 1971 Super is a worm-and-roller type which was used through 1974 ('71 and early '72 featured different internal bushings but it doesn't affect the overall function). Because of the necessary close tolerances, rebuilding a steering box isn't totally impossible, just really difficult. If yours is anything close to ours, you'll replace it; however, in Chapter 6, we'll show you how to rebuild one.

We started this chapter by removing the front fenders. You don't have to, but we wanted to show you the clearest view of the undercarriage as possible. But you'll most likely have to remove them sooner or later anyway, so now is as good a time as any. Just pull off six bolts from each fender, leaving two at the 11 and 2 o'clock positions. This will make it easy to pull out the fender beading while allowing room for the painter to get in all the cracks (or paint the fender detached—which is recommended).

In the next chapter, we will take apart the strut towers, replace the springs with Topline's Sport Springs and reinstall them. For now, you must go through each piece that came off the car and clean them up. Control arms and steering knuckles are virtually indestructible, but if yours are bent, cracked or damaged in any way, replace them now. For the tie rods and steering damper rod, we plan on replacing all of it, so you can throw out the old ones (of course, never throw out any part until you've found its replacement).

Until recently, steering boxes were either still good and usable or bad and replaceable. See Chapter 6 as rebuild kits are available for the worm-and-roller units.

Don't even think of working on a car without jackstands. Hydraulic jacks have been known to lose pressure without warning and collapse.

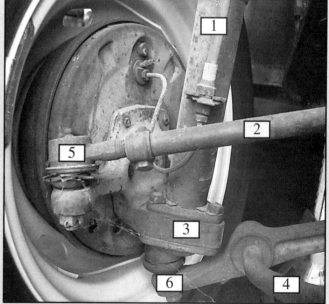

This is the main components of the Super's suspension and the focus of this section: 1. MacPherson strut tower; 2. Left tie rod; 3. Left track control arm; 4. Stabilizer bar; 5. Steering knuckle; 6. Ball joint. All of these parts will be removed and refurbished.

FRONT SUSPENSION DISASSEMBLY

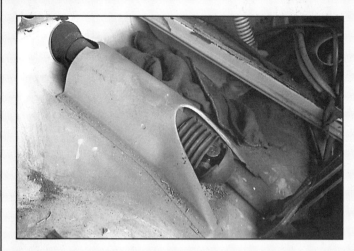

You'll notice that our steering coupler boot is slightly worn, and since they don't make them any more, it is a good idea to take care of yours.

In the spare tire well, the cap on the left hides the locknut for the roller shaft adjusting screw.

Remove the cover above the spare tire well in the trunk. Underneath, you'll find a locknut for the roller shaft adjusting screw. Loosen the lock nut and turn the screw slowly clockwise until you feel the roller contact the worm (if it does). We tightened this down as far as it would possibly go with no positive results. Play must not exceed 15mm, but if yours does (and ours did by a long shot), you should replace the box or have it rebuilt.

Instead of worrying about the steering for now, we moved on to the strut shock absorbers. This clip holds the break hose steady on the strut itself, and if you're working on the driver's side, remove the speedo cable as well.

On the strut shock absorber are three 15mm bolts held in place by three lockplates that keep the bolt from backing out. Flatten the clips with a screwdriver (left) and remove the three bolts (right). You'll need to use plenty of WD-40 here, and perhaps a wire brush to loosen up the dirt.

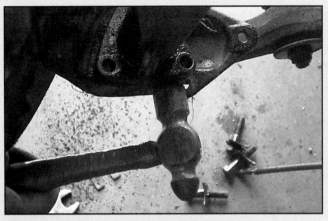

Odds are pretty good that your ball joint and steering knuckle are semipermanently sealed together because of rust and grime. Use a hammer and bang on the steering knuckle to free the two from each other.

A few more good blows on the knuckle and it will slide off of the ball joint.

Once separated, the ball joint will hang down on the control track arm. Be careful not to let the wheel hub crash down on you. Though it is still supported by the tie rod, it can swing out.

Inside the trunk, remove the three 14mm bolts that hold the top of the strut tower in place.

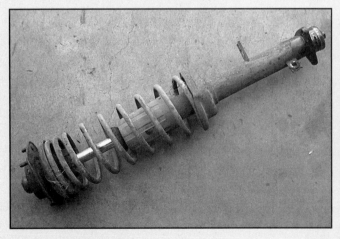
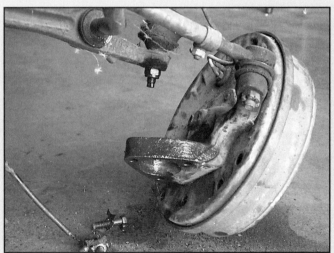

The whole MacPhearson strut system simply pulls out of the car, while the drums are allowed to hang on the tie rods and brake cable.

FRONT SUSPENSION DISASSEMBLY

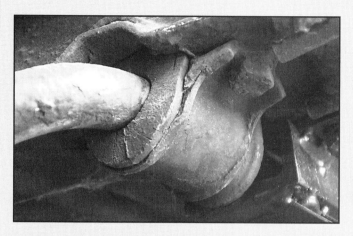

The design of the stabilizer bar and the control arms are such that most of the normal wear and tear associated with a VW's suspension is confined to mostly rubber parts like these frame-mounted bushings around the stabilizer bar.

The two mounting clamps are removed by unbolting the two 13mm bolts on each clamp. Be careful not to break the bolts, as all of the bolts in this part of the car get the most abuse from the elements, and rust may have weakened the metal.

The stabilizer bar is connected to the control arms via these bushings and washers. The end of the bar is threaded to accept the self-locking nut, but first remove the cotter pin.

The bar comes out with a little help from a hammer and awl.

This is a closer shot of the stabilizer bar bushings.

Working our way toward the center of the suspension, this is the connection between the control arm and the ball joint (through '73). For 1974 Supers, the ball joint is pressed into the control arm rather than bolted to the bottom of the strut. For our '71, remove the self-locking nut, and use a fork (or puller if you have one) and separate the arm from the ball joint. Use a lot of WD-40.

THE VOLKSWAGEN SUPER BEETLE HANDBOOK

This is the connection point of the control arm to the frame. These eccentric camber adjusting bolts allow you to adjust the wheel camber by turning them, while allowing the control arm to attach to the frame. Again, lubrication is needed here, as most of these parts haven't been removed from the car before.

With the control arm loose (note the ball joint is still attached in this shot), the drum is only supported by the tie rod and the brake line.

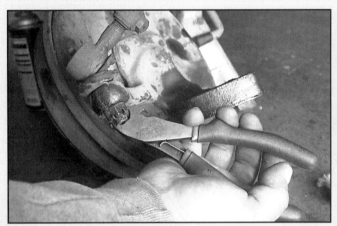
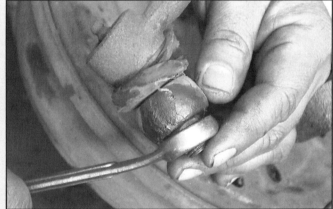

To remove the tie rod from the steering knuckle, remove the cotter pin, unbolt the 19mm bolt.

Using a puller, or in this case, a fork to separate the two. Sometimes a hammer is needed too, and since we are replacing the tie rods, don't worry about stripping the threads.

Remove the cotter pins on the tie rods. Do not reuse them. Always get new cotter pins.

FRONT SUSPENSION DISASSEMBLY

This is the center tie rod, where the left tie rod connects. The center tie rod is connected on either end to the steering box via the drop arm and to the idler arm bracket via the idler arm. All of the tie rod ends are connected to their various points with 19mm self-locking nuts and cotter pins.

The is the working end of the drop arm. In the center is the bolt, washers and bushing sleeve that hold the steering damper to the drop arm, and to the left is the center tie rod connection point.

The steering box is connected to the wheel via two universal joints that help lessen the impact of the wheel and shaft in case of an accident. Disconnect the bottom linkage that connects to the steering box.

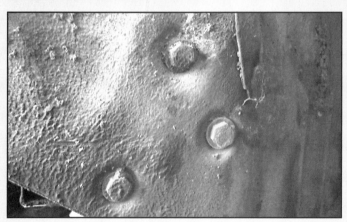

These three 19mm bolts hold the box to the frame. Once these are removed, the steering box merely falls out the bottom, attached to it is still the drop arm.

31

Once the steering damper piston rod is unbolted from the drop arm, remove the cover in the trunk to expose the bolt that holds the rod to the frame head. Replace it if it does not operate back and forth smoothly with uniform resistance.

The idler arm and bracket should be removed as a unit, as it is easier to disassemble outside the car. Similar to the steering box, three 19mm bolts sets the bracket and arm free.

Once all of the suspension pieces are removed from the car, the front end will look pretty blank. Leave it up on jacks in a secure place while you refurbish the suspension parts, readying them to be reinstalled.

FRONT SUSPENSION REBUILD

With the completion of the suspension system, it looks almost better than new. It's slightly lowered and promises to handle better and ride more comfortable.

Once everything has been removed from the front end of your Super, it'll look pretty barren, as there's nothing under there but the frame head and the body. Be sure that you can keep your Super up on blocks while you secure the correct parts you'll need.

The main feature of the front suspension is the MacPhearson strut towers, of course. Since springs are easy to replace (and should be replaced in pairs), we decided to go with Top Line's Sport Spring suspension because it will allow for a stiffer ride while giving the Super's front end a much-needed lowering. Certain things need to be replaced, such as the tie rod ends, all of the bushings, the steering damper and the steering box, as it was determined to be faulty.

Once things are all completely apart, now is the time to check all of the components for wear and damage. Start with the steering knuckle, either with the drum backing still attached or not, it doesn't matter. If you've got a micrometer or vernier calipers, measure the diameter of the outer tapered-roller bearing seat (the smooth area behind the threaded portion of the axle stub). It should be between 17.45 and 17.46mm. Measure the diameter of the inner tapered roller bearing seat. It should be 28.99 to 29.00mm. The diameter of the grease seat seal should be 40.00 to 40.25mm. There should be no distortion of the axle stub, and if there is, you might want to replace it.

It is important to make sure the control arms and center tie rod are straight and free of cracks and unusual wear. Since they are designed in such a way that the replaceable bushings take the brunt of the wear, it is pretty difficult to damage a control arm, unless the car was involved in an accident. If that is the case, you would have trouble realigning the arms with the ball joints and tie rods and the damage would be obvious.

The tie rod ends will be replaced, but the actual rods need to be cleaned and inspected for wear and cracks. remove the ends and roll the rods across a flat surface to check for warping. If they're not straight, replace them. Note: Do not confuse the tie rod ends that are used for front axles with those that are used with strut suspensions. The strut suspension's tie rod end has either a protrusion or an indentation to mark the difference.

The stabilizer bar should be straight and free of any damage. Since it usually gets hit first, sometimes there are small dents on the bar itself. Don't worry too much about these dents unless they affect the integrity of the bar itself. Clean it with a wire brush and repaint.

THE VOLKSWAGEN SUPER BEETLE HANDBOOK

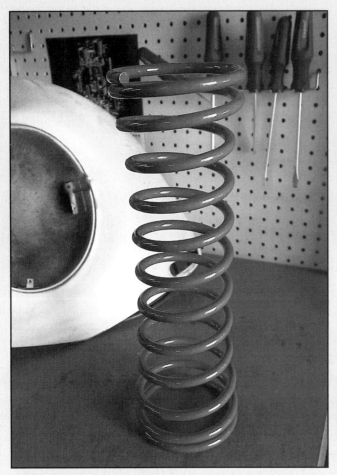

This is Top Line's Sport Spring set made for Super Beetles and Convertibles from 1971 to '79. It is a stronger spring than stock, therefore offering a stiffer ride with a slightly lowered gait.

To disassemble the strut, you must have a proper spring compression tool, otherwise, you could injure yourself severely. Once the spring is secure, pry off the thrust bearing dust cover and remove the self-locking nut from the shock absorber piston rod.

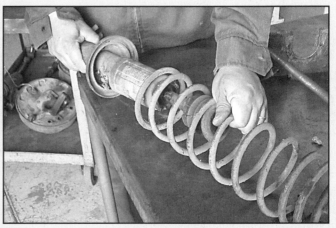

Underneath the rod nut will be the upper bearing plate, the spring seat, the spring and the dust cover. Remove all of these things and write down their order as you disassemble.

The piston inside the strut needs to be removed and replaced. Underneath is a dust cover seat and a locking ring (that you must reuse). The rubber bump stop is damaged here and must be replaced.

With everything removed, the towers can be cleaned and repainted black. Make sure to empty any debris or part fragments out of the strut tube.

FRONT SUSPENSION REBUILD

To reassemble, insert the new shock absorber into the strut tube, the centering cap, the locking nut. Be sure that the spring is properly seated in both spring plates at either end. Unlike the stock spring, the Sport Spring doesn't have a top or a bottom, either end will fit. Tighten the rod nut securely and replace the dust cover. Remove the spring compression tool carefully.

This is the ball joint (upside-down). Since we didn't want to bother with checking whether the ball joint had any play or not (since it is 30-years old), we felt it was safer just to replace it.

The ball joint attaches to the bottom of the strut tower via the three original bolts. It is supported by the control arms and moved with the tie rods.

The three original 14mm bolts also sandwich the steering knuckle between the ball joint and the tower. At this time it is a good idea to install the whole unit together. It makes it easier in the long run.

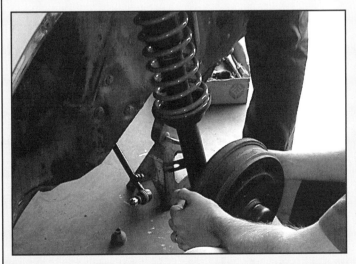

Let the unit hang from the three upper mounting bolts for now. Torque the 15mm bolts to 14 ft.-lbs.

For the control arms, these are the eccentric adjusting bolts and washers. Note the washers' holes are slightly off-center to allow for the adjustment of the camber. Turning the bolt one direction or another pushes the control arm out or pulls it in depending on the desired angle of the wheel.

THE VOLKSWAGEN SUPER BEETLE HANDBOOK

When you buy new bushings for the control arm and the stabilizer bar, it comes with lubrication. Apply it to the inside of the mounting point and the circumference of the urethane bushing.

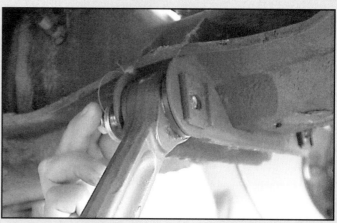

Torque the castellated nut to 22 ft.-lbs. of pressure and install a new cotter pin. Once everything is attached, you'll have to adjust the camber and then retorque this nut to 29 ft.-lbs.

On the other end, thread the ball joint through the control arm and tighten the nut to 29 ft.-lbs. Make sure the ball joint bolt and nut are free of any debris or grease.

The eccentric bolts should be at opposite positions on either side of the frame head to insure matching camber on both wheels. Have it professionally adjusted if you are unsure.

Before installing the sway bar, liberally grease the mounting point on the lower end of the control arm with the supplied grease.

Press the urethane bushing onto both sides of the sway bar. Cover the bushing with grease.

FRONT SUSPENSION REBUILD

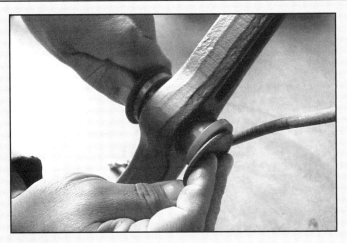

Sandwich the control arm with other half of the bushing, the two washers and the nut. Torque it to 22 ft.-lbs. of pressure and replace the cotter pins.

Line up the sway bar with the front mounting points and lubricate the bar where the bushings will be located.

The mounting clamps attach with two bolts and lock washers each and should be tightened with 14 ft.-lbs. of torque for 1971–'73 cars and 29 ft.-lbs. of torque for '74 and later.

This is the new shiny steering box that will replace the worn out one behind it. You must transfer over the drop arm and universal joint shaft. One or two splines on the drop arm and the roller shaft of the steering box are wider and they must match to install properly.

You must press in the new idler arm bracket bushing and idler arm shaft. The shaft attaches to the idler arm, which connects to the center tie rod on the passenger side of the car.

Once the drop arm is attached to the roller shaft of the steering box (torqued to 72 ft.-lbs.), it is important to peen the lower part of the nut until it meshes with the slots in the roller shaft.

37

THE VOLKSWAGEN SUPER BEETLE HANDBOOK

The universal joint shaft fits snuggly over the steering column splines. The bolt aligns itself with the notch in the splines, and needs to be torqued to roughly 5 ft.-lbs.

The new steering damper is bolted to the frame head through an access hole in the trunk. The hydraulic portion attaches to the top of the drop arm in the mounting point closest to the steering box. Don't forget to replace the bushing at this point (and the sleeve if it is damaged).

The most time-consuming part of the whole job is getting the retaining ring to fit over the tie rod end boots. It takes patience and determination.

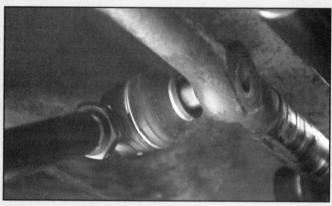

Once the center rod is attached to the drop arm and the idler arm (on opposite sides of the car), the right and left tie rods can be attached with 22 ft.-lbs. of torque. The left tie rod is longer than the right.

The outer tie rod ends attach via a bolts to the steering knuckle. It is at these points you can adjust for toe.

Checking for proper toe-in can be done simply with a tape measure and a friend. Strut front suspension cars are designed to operate with a small amount of toe-in. Measure the distance between two points on the tires, roll the car forward until those two points are in the back and make the same measurement again. The difference at the rear should be less than 5mm greater than the front measurement.

REBUILDING THE STEERING BOX

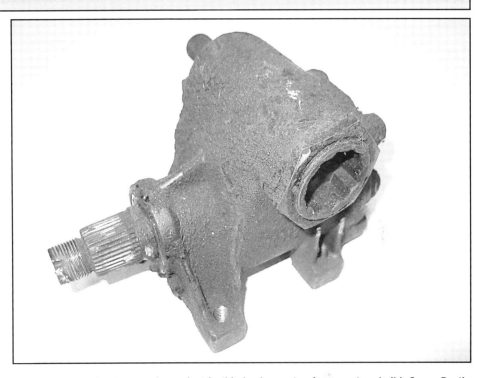

Though the steering box on the project in this book was too far gone to rebuild, Super Beetle specialist Top Line, in Anaheim, Calif., was kind enough to donate a steering box that needed one.

VW used two different steering boxes for the 1971–'74 Super Beetle models and both rebuild kits are available. The 1971 to mid-'72 used one variety, while the mid-'72 to '74 models used another. Starting intermittently at chassis 112-257-5327 for Sedan models and 152-257-5327 for Convertible models, the new steering box's roller shaft end (the part that the Pitman arm fixes to) was decreased to 12mm (formerly 14mm). At the same time, the bearings of the roller unit were shaped differently. The bearings of the late style were small rods, while the early style were spherical. It is important to determine which box you have prior to ordering the rebuild kit. Being that the change was made intermittently at first, the simplest method to verify what style you have is to simply measure the roller shaft end as described earlier.

The most common cause for steering box failure involves wear between two key components, the roller portion of the roller shaft and the worm spindle gear. These two components, over time, tend to wear and thus cause the steering box to become loose and sloppy. A telltale sign is the ever increasing play in the wheel, that empty side-to-side bouncing before the wheel actually moves the tires.

Until now, replacing a worn 1971–'74 Super Beetle steering box meant shelling out roughly $250 just for the rebuilt unit, plus a few hundred dollars to have it installed. Wolfsburg West of Corona, Calif., has the solution for the price-sensitive Super Beetle owner that does not mind getting his/her hands dirty. At around $100, their rebuild kit will restore your worn steering box to like-new condition with surprisingly minimal effort.

The only true specialized tool required for the rebuild process is a torque wrench. The torque wrench must have a range to measure at least 14 ft-lbs. of torque, and a maximum of 50 ft-lbs. Most 3/8" torque wrenches fall into this range. A 48mm socket is also required and should be of the same chuck size as the torque wrench. Both tools can be purchased through any Sears department store. A list of other tools includes a slotted screwdriver, 13mm wrench, socket and ratchet combination with the following sockets: 13mm, 1 1/8" deep socket and 13/16" deep socket. A punch with a 3/16–1/4" flat tip and a pair of large circlip pliers configured to outward expansion. An oil seal puller may also be of help, however one could just as easily use a flat screwdriver to remove the oil seals.

THE VOLKSWAGEN SUPER BEETLE HANDBOOK

Included in the kit is the worm spindle gear, worm spindle seal, steering roller, steering roller support pin, roller shaft seal and the gearbox hosing seal.

With the steering box removed from the car, remove the Pitman arm from the roller shaft. Using a 13mm wrench, remove the locknut from the roller shaft adjusting screw. Next, remove the four bolts that secure the gearbox housing cover. Turn the roller shaft adjusting screw clockwise several rotations; this will free the cover.

With the cover removed, drain the lubricant from the gearbox housing. Rotate the steering worm spindle gear until the roller reaches its centered position. Place a block of wood on the bottom of the roller shaft and drive the shaft free.

Remove the old steering roller from the steering roller shaft by removing the steering roller support pin with a hammer and punch.

Using a 48mm wrench or socket, loosen the steering worm spindle gear adjuster locknut. Once loosened, the worm adjuster can be unscrewed. Generally speaking, the adjuster nut will be able to be loosened by grasping firmly on the threads, then twisting in a counter-clockwise fashion. If it is being stubborn, use a set of circlip pliers configured to outward expansion to remove the worm adjuster.

Remove the old worm spindle gear seal and the roller shaft seal with a seal puller or similar tool. Remove upper worm spindle gear bearing, but be careful not to damage the shim located directly behind the spindle gear seal.

REBUILDING THE STEERING BOX

With everything now removed, thoroughly clean all components with degreaser to remove oil and debris.

Install new roller shaft oil seal using a 1 1/8" deep socket and hammer. Install new worm spindle gear into clean gearbox housing along with the upper bearing. Do not install the new oil seal until after the worm spindle gear has been properly adjusted. Installing the seal prematurely will not permit proper adjustment. Install the lower worm spindle gear bearing and race.

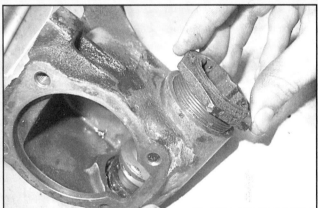

Place a small amount of blue RTV sealing compound onto the threads of the worm adjuster and tighten it lightly using a set of circlip pliers configured to outward expansion. This will press the inner and outer worm spindle gear bearings into their seats. Loosen the worm adjuster and then tighten it slightly until the worm spindle gear feels slightly rough as it is turned. Torque the worm adjuster locknut to 36–43 ft.-lbs. Install new worm spindle gear oil seal using a 13/16" deep socket and hammer.

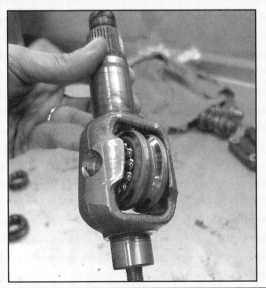

Install new steering roller onto the steering roller shaft using new steering roller support pin with a hammer and punch. Then install the roller shaft onto the gearbox housing cover by turning the adjusting screw through the cover as far as permissible.

THE VOLKSWAGEN SUPER BEETLE HANDBOOK

Install the new gearbox housing cover seal onto the gearbox.

With the roller gear in its centered position, install the roller shaft/gearbox housing cover into the gearbox housing at a right angle. Install the gearbox housing cover bolts and torque to 14–18 ft.-lbs.

With everything back in place, the rebuilt steering box works like a brand new unit.

REPLACING THE PACKAGE TRAY 7

There are plenty of things that can go wrong with a car, especially those as old as an aircooled Volkswagen. Of those things, rust is perhaps the most feared, and people spend countless dollars fighting it. Some would rather crunch a fender than have to do rust repair, but if you've set your mind on restoring/rebuilding your car, then there's nothing you can do but to work through it.

There are several places on this Beetle that have spots of rust, but none more evident than behind the back seat, otherwise known as the package tray. Original window rubber gets old (as does the cheap aftermarket stuff), and when it does, it becomes cracked and porous. After that, there's not much keeping water from coming through the holes. Proper care of your rubber parts is all it takes to avoid this whole situation, but once you can easily reach the starter from inside the car, it's way too late.

There are three ways you can fix this problem: One, pay a lot of money to have someone else do it; Two, tack a plate over the rusty holes and cover it with the rear carpet kit (which is not really a good solution, temporary at best); Or three, cut out the rust and weld in a replacement panel yourself. It's a lot of work, but it can be done by anyone in their driveway with the most basic of tools.

Here's what we have to work with, a rusted, broken scrap of sheet metal, no better at keeping out the weather and engine fumes as a screen door.

Our package tray replacement comes from JP Group, an exhaust and sheet metal company based in Denmark (www.jpgroup.dk). It comes as one piece (Part Number 951088-0), complete with the horizontal firewall and rear engine seal housing. It is a high-quality part and will be a perfect replacement.

You'll need a reciprocating saw, a grinder, a hand drill and a wire wheel, along with a chisel, hammer and a wire brush. Most important, you'll need a welder, much like the Hobart Handler 135 model, specifically designed for 110 home and light shop use. We're using 35-gauge wire for good sheet metal. And, you'll need a good friend if you want to do a quality job. My friend, Chris Miller, was nice enough to stop by and help with this chapter.

Start by laying down a tarp on your driveway. Small rusted particles that fall onto your driveway become small rusted stains, which take a long time to pressure wash off. Since we're working with welding equipment, there are several safety measures you'll need to realize. First off, keep a fire extinguisher nearby. Second, protect your eyes with a good quality welder's hood, and third, protect your hands with thick welder's gloves.

Our replacement piece is a complete tray, including the firewall and engine seal lip. Note the fire extinguisher nearby, as well as our MIG welder.

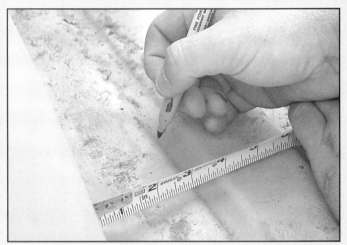

Start by measuring how much of the tray you'll need to cut away. You'll want to make the cuts large enough to include all the rusted parts, but even enough to make easy welds later.

The lower portion (where the seatbelts bolt to the body) is spot welded approximately 40 times, every half inch all along the body. Instead of drilling out each of those welds, we decided to simply pry apart the two pieces, and since we aren't keeping it, it doesn't matter how pretty it looks.

It doesn't look good, but the spot welds are gone and all of our pieces are cut out.

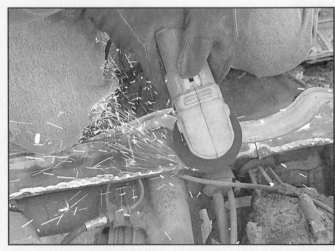

A coarse circular grinder works wonders here, shaving down burrs and removing any spot welds that are still left.

REPLACING THE PACKAGE TRAY

After this, wire wheel all of the body sealer away from the area you need to weld. Welders need a solid metal-on-metal finish to work best, and this is the quickest way to do it. Once this is done, turn your attention to the new sheet metal.

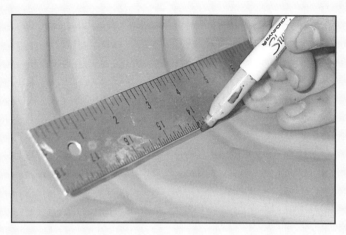

It is best to measure a half-inch larger than you think you'll need. Once you cut it, there's no turning back. If you're wrong, you'll either have to get another piece or do some trick welding.

Like we stated earlier, we only planned to replace the lower half the tray, so we cut it on our pre-measured point, approximately in half. Sure a plasma cutter would be best here, but really, who has one of those in their garage?

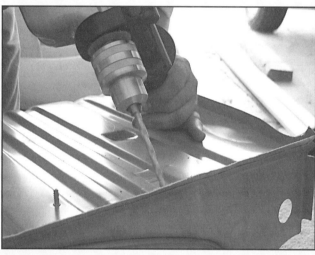

Also, since we aren't planning to replace the lower firewall and engine seal lip, we drilled out the 12 spot welds and separated the two pieces...anyone need a new firewall piece?

An important step in the process is to make sure things fit as you go along. Even though you know it won't fit, set it in the rear of the car and see how you're doing. Since we cut things a little on the large side, we've got some room for adjustment.

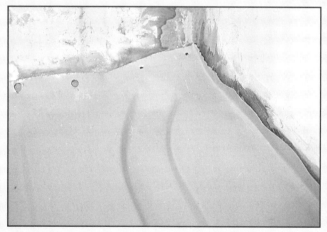

Since the car had been hit on the rear passenger side, there was a small indentation that needed to be accounted for. This is the left rear corner (driver's side) that we used as a guide point. If this was square, everything would be measured from it.

THE VOLKSWAGEN SUPER BEETLE HANDBOOK

After a few final adjustments and test fits, the piece fits as flush as possible against the far inner fender wells. If you're happy with the fit, warm up the MIG welder.

Start welding in both lower corners, those nearest the tunnel and transmission inspection plate. By spot welding from the front to the back, you can make sure each portion is flat and flush without any waves.

To hold down some trouble spots, use longer welds like these. It is a good idea to keep a water bottle handy to cool welds and to keep from burning yourself as you go.

After the spot welds, everything is looking straight and solid. Next thing to do is to run a bead of welding around the entire perimeter of the new sheet metal.

REPLACING THE PACKAGE TRAY

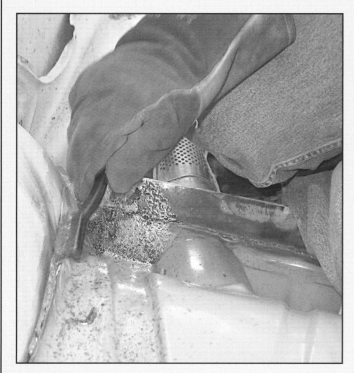

A wire brush removes some of the slag and keeps the debris from contaminating the weld.

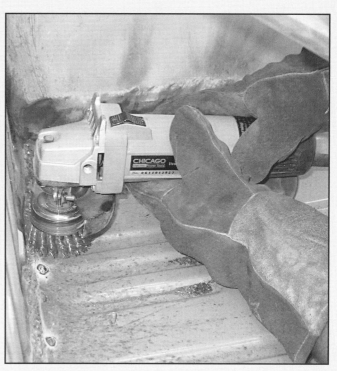

After the welding, wire brush and grind the weld scars clean and smooth.

You'll notice the addition of the rear carpet kit rail. It is simple as drilling out the 12 spot welds and rewelding it to the new location, 14 1/2 inches from the firewall.

Here's the final product. Use a sealing primer to cover all exposed metal, and don't worry about the small weld scars as no one will ever see them.

8 REBUILDING THE BRAKES

A lot has to happen when you step on the brake pedal in order for your car to come to a safe stop. Levers, pneumatic pressure, cables, cylinders, springs and friction all play critical rolls in the process, and we place a lot of blind faith in this system, considering how important it is. Braking is indeed a life-or-death science and should be treated as such. That's why brake fade, ever-increasing pedal travel, dragging brakes, or ones that lock up while traveling are serious symptoms of a faulty system. If you have any of these problems, it is time for an overhaul.

There are several major components that make up the mechanical braking system for Super Beetles. The master cylinder is actuated by the brake pedal and is connected to the cylinders at each wheel through the hydraulic brake lines. The two pistons in the master cylinder operate the front and back brakes respectively, and the fluids are from the reservoir located in the trunk and connected to the master cylinder via two hoses. The wheel cylinders have hydraulically operated opposed pistons that press the two brake shoes onto the drums. In addition to this, the parking brake is operated by a cable and is connected to the rear brakes only.

If you have any doubt about the operation of any of the brakes' components, replace them.

Shown here are the various parts associated with the rear brakes. The front setup is similar, except for the cable and lever for the parking brake. 1. wheel cylinder; 2. brake shoe; 3. upper and lower return springs; 4. adjuster screw; 5. shoe retaining springs, clips and cups; 6. connecting link.

Another point of interest are the drums themselves, and should be thoroughly inspected before deciding whether they can be reused again. Any tapering, scoring, or other unusual wear should be noted and considered. If you have calipers, it is a good idea to measure the inside diameters of the drums. If those measurements are greater than 231.5mm for the rear and 249.5mm for the front (the permissible wear limit), they'll have to be replaced.

All you need are a few metric wrenches, screwdrivers, a pair of plies and a breaker bar to do a complete brake rebuild.

BLEEDING BRAKES

Once you've completed the procedures in the pages that follow, you'll need to bleed the brakes. Bleeding the brakes removes air from the lines, and must be done if the wheel or master cylinders were replaced. The easiest method requires two people. Have a

REBUILDING THE BRAKES

partner sit in the car and pump the brakes while you move from wheel to wheel, starting with the front right wheel. Find the bleeder valve, attach a hose over the valve and submerge the other end in a jar of clean brake fluid. Open the valve a half turn and have your partner depress the brake to the floor and hold it. Close the valve, return the pedal to the upright position and repeat the process until there are no bubbles. Do the remaining wheels in this order: left front, right rear and left rear. Refill the reservoir after bleeding each wheel and don't let it go dry.

Before buttoning everything up, double check to make sure all nuts and bolts are properly torqued and that all of the components are in the correct place. If you have a motor in your car, take it out for a test drive (in a safe area, of course). Your brakes shouldn't be uneven, pull any one direction, squeak excessively or pulsate erratically.

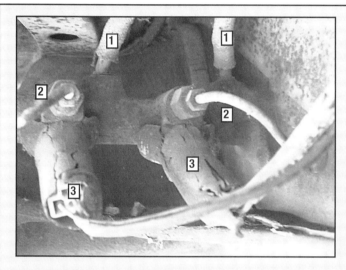

Here is the master cylinder as it was for 32 years. For 1971 and '72 four-wheel-drum systems on Supers, there are residual pressure valves between the master cylinder outlets (hexagon shaped bolts) and the pressure lines that lead to the wheel cylinders. 1973 and later Supers (and those with disc brakes) have restriction drillings instead of the valves, and you must not interchange the two (to tell the difference there's a V-notch on the mounting flange of the cylinder with restriction drilled cylinder). There are three important things in the photo to note: 1) The two hoses that connect to the reservoir are at the top of the picture; 2) The two pressure hoses that head to the wheel cylinders; and 3) The brake and warning light switches connect to the master cylinder and the fuse box/electrical system.

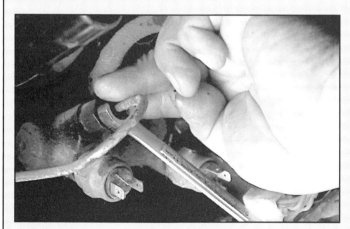

Removing the master cylinder is as easy as disconnecting the hoses to the reservoir, unplugging the brake light wires and unbolting the pressure lines to the wheel cylinders. If you're not planning on replacing the lines, make sure to protect them with dust caps.

From inside the car, disconnect the pedal assembly from the master cylinder (through the pushrod) by unclipping the circlip and prying off the pedal-return spring. Then just unbolt the two bolts and the master cylinder falls free.

Before jacking up the car, we replaced the old master cylinder with the new one by basically doing the above two steps in reverse. We also replaced the brake/warning light switches, and the reservoir elbows that came with the new cylinder.

Jack up the car (whatever end you decide to work on first) and make sure you block the opposite wheels and set jackstands underneath. Remove the tires and place them under the frame of the car, pull off the dust cap and remove the nut, thrust washer and outer tapered-bearing (which we will replace, of course). The drum pulls off easily and underneath is the working end of the brake system.

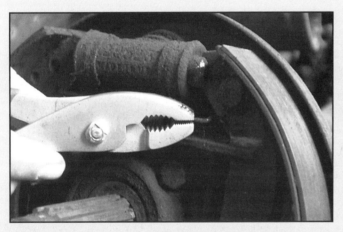
Though this isn't the greatest picture, we are removing the insides, starting with the upper and lower return springs.

After some easy work, brake parts will begin to come off. To remove the shoe retaining springs, push in the cups and rotate the unit 90 degrees. With those out, the shoes fall out.

Here is a closer shot of the lower portion of the backing plate. The adjust screws and the gear-shaped adjuster that fits into the anchor block (shown) have already been removed. The cable is the parking brake. You don't need to do anything with it.

This is the pressure line as it connects to the back of the wheel cylinder. The threads above this line are the connection point for the cylinder to the backing plate.

REBUILDING THE BRAKES

Though this seems worn, the drum is actually just fine, though I know many thousands of miles have been driven on it. If you have any doubt (or don't have the means to properly measure it) take it to a professional.

Replacing the old wheel cylinder with a new one is as easy as bolting it to the backing plate. Even if the old cylinder seems fine, they don't cost too much, and it is safer to replace them while the wheel is apart. Torque to 18 ft-lbs.

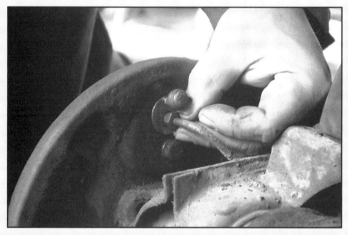

Fitting the pressure line through the backing plate to be mated with the wheel cylinders, you can finger tighten the line.

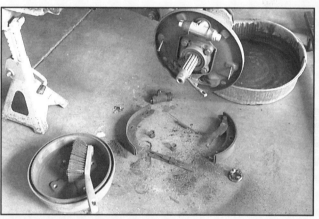

This is where we stand so far. We've kept a brush handy to remove any brake dust that has accumulated on the new parts. As well, it is a good idea to give everything a fresh coat of paint, especially the outer surface of the drum.

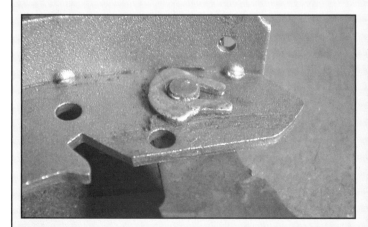

Before replacing the shoes and pads, connect the parking brake lever to the shoe with the pivoting pin and clip. It's easier to connect it now rather than after you've attached it to the backing plate.

After the shoes have been replaced (with the retaining springs) attach the connecting link and upper and lower springs (shown here).

THE VOLKSWAGEN SUPER BEETLE HANDBOOK

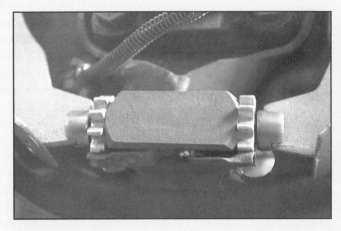

These are the adjuster screws as they plug into the anchor block. The gear teeth can be used to tighten the shoes closer to the drums from the rear of the backing plate so you don't have to remove the wheel and drum.

At the base of the right shoe is where the parking brake lever connects to the cable. It is as easy as slipping the cable over the lever.

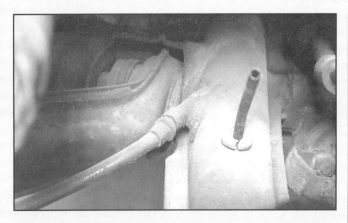

This is the pressure hose that connects to the hard line through the tunnel (so they're not subject to damage). These hoses need to be replaced and it is as simple as unbolting the old ones and replacing them with the new lines (torque to 11 to 14 ft-lbs.).

The hard lines are secured via clips to the frame. These prevent them from moving excessively at their connection point to the hoses.

The hard lines are connected to the soft hoses in the rear at this point. Nice clean new hoses attached to an old body part. After bleeding the lines, the system is done and complete.

Replace the drums, roller bearing and nut. Put the wheels back on and lower the car. Torque the castellated nut to 253 ft-lbs. and the wheel nuts to 95. Make sure you do these torques when the car is on the ground, as there is enough force to push the car off the jack stands. Install a new cotter pin and prepare to bleed.

REPLACING THE SPRING PLATE BUSHING

We've concentrated most of the first few chapters on the suspension and brakes, as it is an important step before tackling the rest of the car. Of course, like most of you, we can't afford nor do we have the time for a 100-point pan-off restoration that will convert this car to 110 percent better than original. Our goals are simple: return this Super Beetle to the road so it is safe to drive and nice to look at. We don't intend to win any awards with it, but in the end, it will be a quality job well done.

What we are going to accomplish in this chapter is relatively straightforward and easy provided you have the proper tools. However, if you've never done it before, expect it to take the better part of the day to accomplish, provided you take your time and make sure everything goes smoothly. Remember that we are working with suspension parts, therefore they are potentially dangerous if not well supported. You'll need the usual assortment of wrenches and sockets you'd find in a nicely-stocked garage, along with a floor jack, stands and a tensioner like the one you'll see in this chapter. They can be found at any tool store.

Underneath the torsion bar cover is a set of rubber bushings that, over time, wear out and cause rear sagging. We are going to replace these rubber bushings with much stronger and durable urethane.

Each rear wheel is independently sprung with trailing spring plates, torsion bars and shocks. The rear suspension is centered around the use of torsion bars that absorb the shocks from the road. The function of the spring plate is to support the rear road wheels and provide a pivot point for the torsion bars on which to absorb the bumps. The rubber bushings in the plate tend to wear out. Replacing them with urethane bushings is a big improvement.

The whole process does seem complicated, but if you take it slowly, you should be able to accomplish this task yourself. Most of all, be careful. The torsion bars are heavily preloaded, which can be dangerous if it releases suddenly.

53

THE VOLKSWAGEN SUPER BEETLE HANDBOOK

Start by jacking up the rear of the car and blocking the front wheels so they don't have the chance to move.

This is the rusted mess of the spring plate and the diagonal arm, and the focus of most of the day's work.

As you can see, the top of the shock, at the mounting point, is well secured by rust and dirt. Use WD-40 to cut through most of it before turning to the wrench.

Remove both the top and bottom bolt for the shock absorber. You can discard the two shocks because they will be replaced.

Use a scribe or screwdriver and mark the position of the diagonal arm where it bolts to the spring plate (see arrow). This will give you a reference point to reattach the spring plate into the elongated holes. The holes are elongated so that you can adjust toe-in or -out for rear alignment.

Remove the four bolts that are holding the diagonal arm to the spring plate.

REPLACING THE SPRING PLATE BUSHING

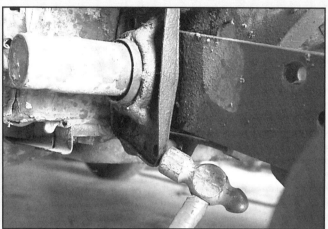

Remove the four bolts that secure the spring-plate hub flange and set it aside. You may need to tap the flange out with a mallet. Warning: The spring plate is resting on the lower stop and is under extreme load. Be careful that this doesn't slip off, as it can be very dangerous.

Once free from the car, the spring-plate hub flange needs to be cleaned up and repaired/replaced if damaged.

At this point, you can do one of two things. Either pry the spring plate off the stop and lower it down slowly (like we did here), or use a tensioner and lift up the spring plate, force a screwdriver under the plate and lower it back down so it slides over the screwdriver. Be careful that the plate doesn't bind on the screwdriver.

This is where the spring plate rests after removing it from the stop.

With the spring plate off the stop and unloaded (but still connected to the torsion bar, make a scribe for a point of reference and then pry off the cover.

THE VOLKSWAGEN SUPER BEETLE HANDBOOK

Underneath is the outer bushing, the torsion bar and the inner bushing. It is a good idea to examine the torsion bar for any spline damage and signs of rusting. Replace it if you have any questions, but if not, make sure there are no spots of exposed metal that will eventually rust.

Place the new bushings into the torsion holes, and make sure you use plenty of the supplied grease.

With the tensioner, load the spring plate by lifting it back up to the scribed mark and the lower stop.

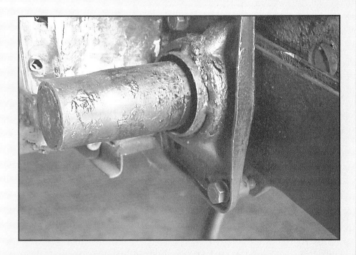

Replace the hub cover and torque to 80 ft.-lbs. of torque.

Lower the wheel assembly (left) and replace the bolts (right). Torque them to 87 ft.-lbs.

REPLACING THE SPRING PLATE BUSHING

This is a nice overall shot of the car's underneath, showing us that we've got a lot of cleaning still to do before the VW is considered finished.

Replacing the shocks is as easy as removing them. Tighten the nuts to 43 ft.-lbs. of torque.

10 REMOVING UNDERCOATING

The undercoating on our Beetle is 35 years old, but it's still bulletproof. Of course, that's why it was added; to protect the undercarriage of the car from the elements, weather, road debris and anything else it happens to run over. In addition, undercoating acts as a noise deadener, suppressing sound waves before they can travel through the metal of the fender and into the cockpit. So, it's pretty important stuff to have, and if it's missing, you'll notice it the first time you head down the road.

If you're redoing a Volkswagen and you'd like to make it as clean and nice as you can on top as well as underneath, then you're going to have to deal with it at some point. Underneath our Super Beetle is covered with old, dirty and spotty undercoating. The rear clip was replaced at some point, but the previous owner neglected to put on some undercoat, so there is some light surface rust in this area that needs to be removed as well. By the end of the day, our goal is to have a shiny, clean undercarriage, which is a challenge.

There are chemicals, abrasives, sandpapers, wire wheels that require a lot of elbow grease and take the better part of a weekend. But what methods work best and which ones don't work at all? Well, we decided to try several of them, and in this chapter, we'll show you what worked best for us.

We bought a host of wire wheels, abrasive flap wheels, wire cup brushes, nylon wheels and a paint and rust stripper. We also picked up a paint and epoxy remover, a fast-acting auto stripper and a Bernz-O-Matic propane hand torch. To power some of these wheels, we bought an angle grinder and a power drill. And because we'll be ending up with bare metal, we need to protect it with a primer.

Since we know a couple of these methods will be messy, and to keep the driveway relatively clean and unstained, we covered it with a tarp and put some drop cloths around the wheels, tires and suspension. We rolled the car onto the tarp and got busy with our first method.

But first, a word on safety. Drills, angle grinders and caustic chemicals are bad when taken internally. Wear gloves (not the lightweight latex ones, but heavy-duty rubber), eye protection and if you're sensitive to sounds, ear plugs. Keep a fire extinguisher handy too.

So, what method is best? Well, unfortunately, there is no one perfect way to remove the material. The chemical is messy, the wire wheel gets gummed up and the torch doesn't get it all. So, what did we do? Combine a couple of them and you're in business. What we liked best was removing all of the major undercoating using several coats of the Auto Strip, cleaning and drying the remaining area and then using the angle grinder with a steel wire wheel. What was left was bare metal ready for primer, paint and reundercoating.

Here's what we have to start with, four fender wells covered with the rubberized tar. Sprayed on about an eighth of an inch, it can withstand most of what you can throw at it.

REMOVING UNDERCOATING

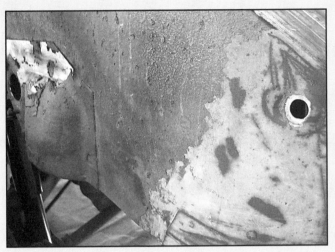

On the right, the undercoating has been removed because of some prior bodywork on the rear apron. What is left is a lightly rusted surface that will need to be tended to.

Over the course of the last 35 years, dirty and grime has built up in the nooks and crannies of the suspension mount points and the body-to-pan connections.

Here's what we have at our disposal, an array of chemicals and drill attachments. Not shown is the angle grinder and its wire wheel.

Up first is Auto Strip from Klean Strip, a paint remover touted as being able to "remove all finishes," so we thought it wouldn't hurt to try. On the back of the label, there's all kinds of warnings, including the one that says not to put it on your skin—yes, it burns.

It sprays on evenly and goes to work immediately, bubbling and sending off an incredibly horrible smell, which is most likely unhealthy, so wear a good respirator mask, like you'd find in an autobody shop.

After the recommended 15–20 minutes curing time, a plastic scraper easily removes most of the undercoating. Aren't you glad you laid down plastic to catch it all? And if you're only wearing latex gloves, you'll soon notice the Auto Strip eating through the latex and burning your hands.

However, not all of it was removed, as you can see. Some patches of undercoating wouldn't budge.

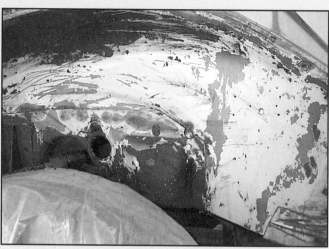
After a second coat of Auto Strip and the use of a metal scraper, the remaining undercoating came off easily along with patches of paint.

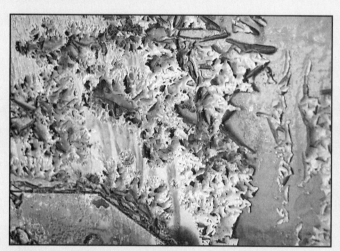
We added another coat to see how good it would remove paint, and after the 20 minutes, and it immediately started to bubble up.

This is 3M's Heavy Duty Stripping Tool, which they suggest to use with chemical paint strippers.

As you can see, it doesn't work too well. Sure, we were allowed a couple of good swipes with the stripping tool before the undercoating gummed up the fibers. The area we did get to use it on looked good, but it couldn't possibly hold up under the gunk of four fenders.

In the meantime, we took a look at the front fender and got out the power tools. The undercoating here isn't as thick as it is in the rear, but it still needs to be removed.

REMOVING UNDERCOATING

Here is our collection of attachments from the hardware store. We grabbed everything that looked like it was designed to remove material from metal.

We started with 3M's large area Automotive Paint and Rust Stripper, an attachment for a power drill that is pretty abrasive.

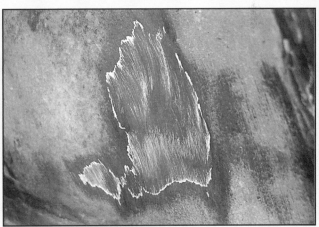

It worked for a while, but soon, the wheel heated up the undercoating and it quickly gummed it up to the point where it no longer worked effectively. All that was removed was a small patch.

Next up is Vermont American's Coarse Wire Cup Brush, with brass-coated, hardened steel wires. It worked well, but it was time-consuming. A small patch took nearly 10 minutes, so you can imagine how long the rest of the car would take.

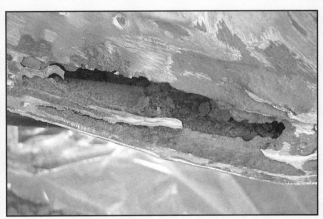

The undercoating may actually hide body damage that was previously undetected, as was the case here. Unfortunately, this hole is very close to the pan and will have to be taken care of before we finish the paint and body.

61

THE VOLKSWAGEN SUPER BEETLE HANDBOOK

This 60-grit Abrasive Flap Wheel from Columbian fits on any type of power drill and has an 23,000 rpm limit. Sixty is a nice grit for removing paint, so we gave it a try.

As you can see, this is all we could remove with the Abrasive Flap Wheel, so off it goes to the failed pile.

Next up is this three-inch Nylon Wheel also from Columbian. Its 120-grit rating did absolutely nothing. It wouldn't even scratch the regular body paint.

This is Jasco's Premium Paint and Epoxy Remover, which is used by professionals as the can says to remove most anything from anything. We had some trouble applying it to a vertical surface, and it would have worked better had we been able to keep the remover in contact with the undercoating.

However, in a short time, this is the results, which is quite impressive considering how little time we spent applying the single coat. Three coats later and the fenderwell was almost clean.

Okay, time for the big guns. This is a 4.5-inch angle grinder from Ryobi, one of the least expensive tools on the market. With the supplied brass-coated steel wire wheel, we made short work of any remaining undercoating.

REMOVING UNDERCOATING

Twenty minutes later, the fenderwell (that was first treated with the Auto Strip was free and clear of not only all of the undercoating, but the original paint and primer underneath. It was a lot of work to get it to look like this, but definitely worth the effort.

Okay, we heard about this method from a few experienced body men, but we weren't sure how good it would work. This is a Bernz-O-Matic propane hand torch that can be found at any home supply store.

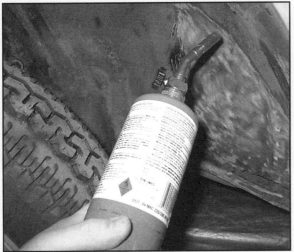

Here's how it works: Heat up the surface of the undercoating and you'll also heat the metal underneath. The metal will get hotter quicker, causing the undercoating to gel. When it starts to bubble, scrap it away with a putty knife. The downside, it's smells bad and is probably toxic.

After a few short minutes, most of the major undercoating has been peeled off. At last! Here is the final result of our troubles. By using a combination of methods and a coat of primer, this is what the finished product looks like.

11 BASIC BODYWORK

In this chapter we will tackle the ins and outs of bodywork, specifically bodywork in your own driveway or garage. It is a lot easier than you think, and the next time you collect a dent or a deep scratch in your car, you'll feel slightly more confident to remove and repair it without writing a fat check to your local body shop.

The first step is making sure you have the right set of tools. Tools specific to body repair are crucial and, because a claw hammer and a few scraps of sandpaper aren't going to cut it, you're going to have to invest in a few items.

First is a dent puller kit, and the second tool is slightly more serious, Eastwood's Magna Spot Studwelder, complete with a heavy-duty slide hammer, T-puller and a collection of pins, rivets and electrodes. Third, we'll finish up with a set of dollies and hammers to smooth out the rough spots, so hopefully we can avoid using Bondo (though most projects will need some amount of filler).

The crucial first step to any bodywork is keeping the surface clean. Move your car into a shaded area or set up an EZ-Up if not only to keep you out of the sun, but to keep the temperature of the car to a minimum. We left a fender out in the sun for a couple of hours of 100-degree heat and the surface temperature of the fender peeked at 172-degrees. That's not something you can easily work on.

Wanting to clean up the front end a little, we sanded smooth most of the rusty spots on the hood and front apron and gave them a protective coat of primer.

After exposing the bare metal of the area to be repaired by removing the paint and primers, clean the area thoroughly with a wax and grease remover. Keep it handy as you'll want to maintain that level of cleanliness until you apply the first coat of primer.

Before we begin the process, let's discuss safety. Remember to use gloves and glasses when working with chemicals and paints, and since we'll be performing a kind of welding, keep a fire extinguisher handy just in case.

With our easy dents, we lucked out in that our methods worked quite quickly. Remember that the contour of the dolly must fit the contour of the damaged area. If the wrong surface hits the panel, like a sharp edge of the dolly, for example, you will cause further damage to the panel. Start with light blows and gradually increase the force of your blows to raise the damage. It is always better to use many well-placed smaller blows than a couple of hard ones. This will give you better control. As you hit the panel, the dolly tends to rebound slightly (and the panel will rebound on its own). This creates a secondary lifting, and to combat this, release pressure as soon as the hammer hits the panel or use a slightly larger dolly.

BASIC BODYWORK

This is the best shot we could get of the first dent, a fist-sized bend technically called a "low spot," a recession below the surrounding surface, on the passenger side of the hood, conveniently over the hood's structural frame underneath. The only way out is to pull it, as the space underneath is a little too tight for spoons (used to pry out damage in tight spots) or dollies.

The second dent is this crease on the front apron formed from hitting a boulder 15 years ago. Since this is another low spot, more like a valley, we'll have to raise out the bend.

The first step for us is to clear the area of the paint and primer. Though this isn't totally necessary, we wanted to see how the metal was underneath the affected areas. We used an easily-found paint stripping chemical and waiting until the stripper did its work.

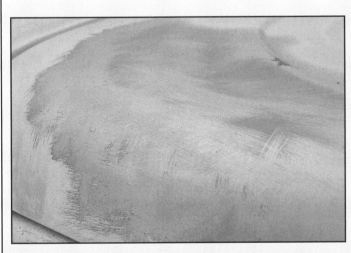

Here are the two bends without paint. Bare metal looks nice and clean, but make sure so with the wax and grease cleaner. Dry it with a lint-free towel and refrain from touching the metal excessively.

THE VOLKSWAGEN SUPER BEETLE HANDBOOK

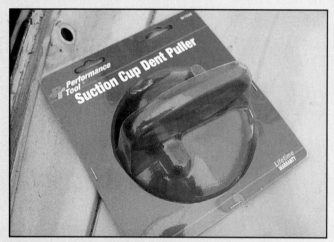

Our first thought was that we could simply pull out the dent with an over-the-counter suction cup puller, but because of the curves of the hood and apron, it wouldn't create a suction. Perhaps if the dent had been on the roof or the door, this would have worked.

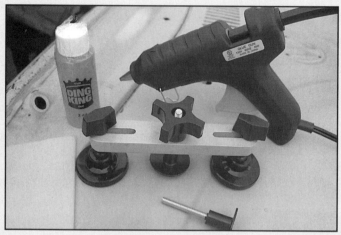

Next up is the "As-Seen-On-TV" dent puller from the Ding King. Simply clean the area again with the supplied cleaner (in the bottle) and apply the glue to the metal.

Once the glue sets up for a few minutes, press down the center piece and twist it back up. Does it work? Yes and no. It pulled out the dent slightly, but there is so much spring back (the ability of metal to return to its original shape, dented or otherwise) that the dent wouldn't come out all the way.

Time to bring in the big guns. This is Eastwood's Magna Spot Studwelder, with a heavy-duty slide hammer, T-puller and a collection of pins, rivets and electrodes. Though it's roughly $300 for the kit, it is surely worth it if you've got several dents on several cars (compare that with the bodyman's bill).

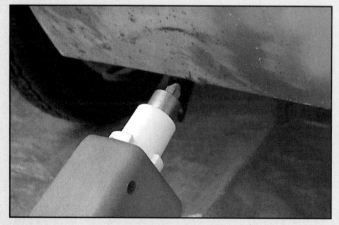

The electrode (the small angled tip) comes in several shapes and sizes, but we used 2mm studs and this is the appropriate electrode. The cylinder surrounding the electrode is the insulator that covers the electrode and the pin.

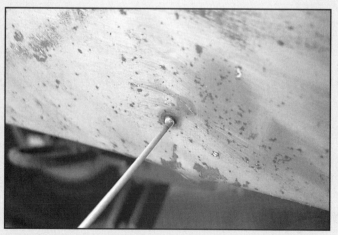

Simply place the electrode at the low point of the dent, press in the insulator and push the button for approximately one second. The pin is welded to the panel.

BASIC BODYWORK

To test the waters and find out how resilient our dent is, we started with the supplied "T" puller, a compact tool that clamps onto the pin and is used for lightweight dents. It prevents over-pull, so you don't end up with a high spot instead of an even panel.

With the results of the "T" puller not exactly what we needed, we attached the slide hammer, a polished steel bar with a rubber grip handle and a cast-iron slide. According to the manufacturer, it provides up to 500 lbs of pulling strength, definitely plenty of power.

It worked so well on the apron that we used the Studwelder on the hood dent. We didn't just use one pin, however, we needed approximately four on the hood and six on the apron.

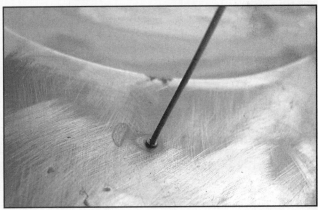

This is to show you that you don't have to weld them in straight and that they can be welded to whatever angle you might need to pull the dent smooth.

After you are done with the pins, simply grind them off and smooth with any angle grinder. Be careful not to cut into the metal or you'll be back to square one.

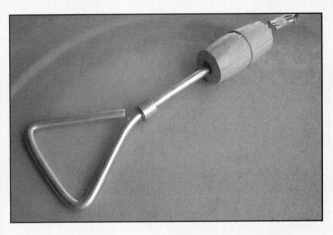

If shelling out $300 for a stud welder isn't in your project's budget, many local automotive stores offer knockdown versions of the same thing. The drawback is that you need to screw the pins into the metal, leaving quarter-inch holes that need to be filled. This slide puller was approximately $8.

67

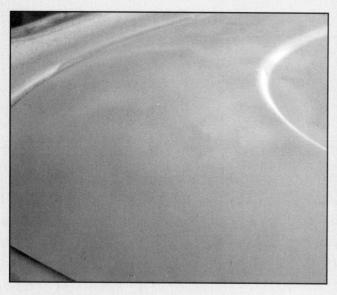 

After a few of hours of pulling and welding and grinding and pulling, this is the results (after a coat of primer). It isn't perfect, but after some sanding and perhaps a thin layer of filler, we'll be ready for paint.

For the apron, we broke out the hammers and dollies to "bump" out the dent. Remove all dirt, undercoating and paint/primer from the backside of the panel as well before beginning. The reason there are so many different types of dollies is so you can match up the shape of the panel with the surface of the dolly. One side of the hammer is rounded for bumping concave surfaces while the flat side is for high spots.

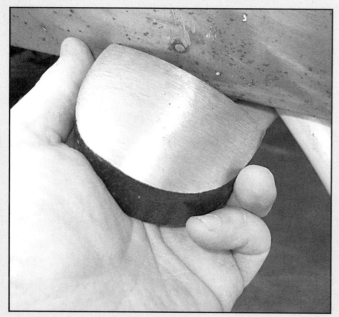 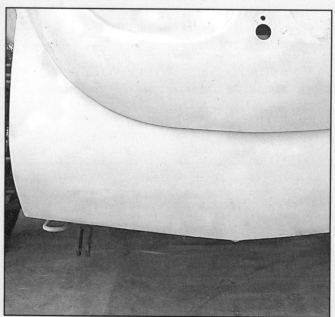

There are two methods to bump dents: off-dolly and on-dolly. The hammer-off-dolly is used to raise low spots and lower high spots simultaneously. The hammer-on-dolly is a method used to apply a concentrated force on a small area. Place the dolly against the top of the damaged spot. The hammer blow exerts a pinching force on the metal, and a small amount of metal is crushed and flattened between the hammer and the dolly. Each blow should overlap the last and no two blows should be on top of each other. A correct hit will create a high-pitched "ping" noise, while a miss will give you a dull or dead sound.

After an hour of properly placed hammer blows (always starting with light blows to correct for aim and tension), the panel had worked its way out. Some spots reversed and became high areas that needed to be battered back down.

REMOVING WINDOW GLASS & DOOR LOCKS

12

In the last chapter, we did some basic bodywork but we still have a lot to do before we go to the paint shop. The body needs to be separated from the pan, and we still need to deal with that rust in the fenderwells. But as I stood in front of the Beetle, thinking I was almost ready for paint, my wife pointed out that the windows were still in it, along with door locks. It seems that in my haste, I had forgotten completely about the door hardware, windows and regulators.

So, in addition to those leftover parts, I decided to do my paint guy a favor by removing the decklid and hood, and along with the fenders, get them media blasted and primered.

There are very few tricks involved with removing the hood and decklid. Eight bolts and a spring are all that hold the two parts to the rest of the car. For the hood, use a 13mm socket and remove the two inner bolts first, then the outer ones. This way, the weight of the hood rests on the outer bolts and the hinge pivot bolt. Leave the hinge in as it can be painted on the car.

The decklid is easy still, and

These Kamax 5.8 metric bolts have been untouched since they were first installed 35 years ago. Since the front end of this Super has never been removed, the bolts are perfectly lined up, pulled far to the front as they should be. If yours has been adjusted in a previous repair, make sure to score the metal so you can line them up again after paintwork.

there's no right order to remove the spring or the bolts. We chose first to unclip the spring with a pair of pliers before we unbolted the 10mm attachments.

With these two parts (and the four fenders) off to our friendly sandblaster, the overall car is seriously lacking the rounded character of a typical Super Beetle, but nonetheless looks like progress.

Follow along as we tackle the windows and doors on our never-ending quest to rebuild/restore a once-beautiful Super Beetle.

THE VOLKSWAGEN SUPER BEETLE HANDBOOK

Here is our 13mm ratchet hard at work removing the rearward bolts first so the hood can rest on the front bolts and the hinge. If you've got good balance, you can remove the hood alone, but it is best with a friend holding the sheet metal while you unbolt the remaining bolt.

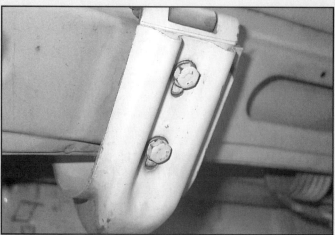

The same holds true for our decklid hinges, as they've never been affected by an accident and are still in their original location. Again, if yours isn't, score them to retain the position so your lid will line up again later.

Squeeze the spring together with pliers and pull it free. Make sure you've got a hand on the lid or it will come crashing down on your head, followed, no doubt, by a woven tapestry of obscenities. To keep the lid up, place a screwdriver in the hinge (like in the next photo).

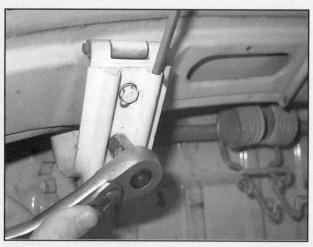

Of course, you can remove the four 10mm bolts first and merely lift the lid (and load spring) off of its bracket. Either way, the job's done, right?

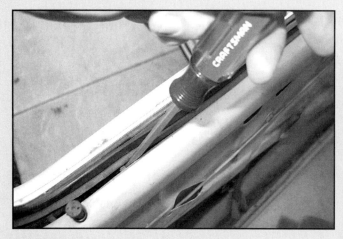

Moving to the doors, the first order of business is to remove the inner scraper, and this is held to the door by four clips, which are easily snapped free.

This 10mm bolt shown here is the lower mount point for the vent window frame.

REMOVING WINDOW GLASS AND DOOR LOCKS

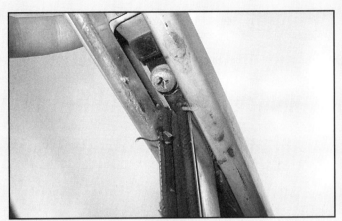

After pulling free some weatherstripping at the top of the vent window frame (on the door window side), you will find this Phillips-head screw. After removing this screw, and with some finagling, the entire vent window assembly pulls out. You may have to wiggle the window from side to side for it to clear.

For the window regulator itself, we'll start with these two 10mm bolts that hold the crankshaft to the door frame.

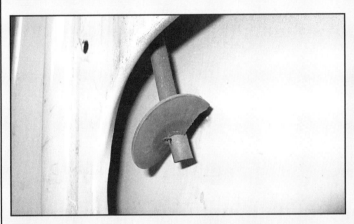

You'll notice that this door has hardly ever been touched, as it is a pretty rare thing to have the original rubber grommet still attached to the regulator. Its function is merely to keep it from rattling in the door (something our driver's door is missing).

These two 10mm bolts and one directly below them are all that's standing between you and an empty door. These two hold the window to the regulator and the one below holds the regulator to the door.

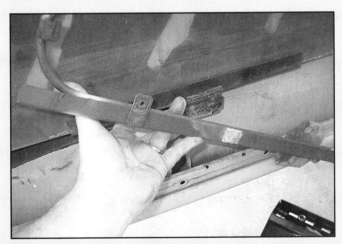

As you can see, once gone, the window and regulator practically fall out of the door, so be prepared.

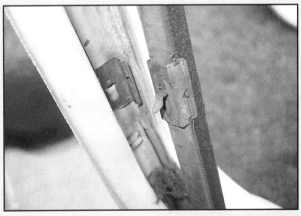

This is a closer look at the retaining clips for the felt channel, specifically along the outer edge of the door. You don't have to keep any of these as they can be had brand new (for example, from West Coast Metric). We tore the felt off at a point just below the windowsill.

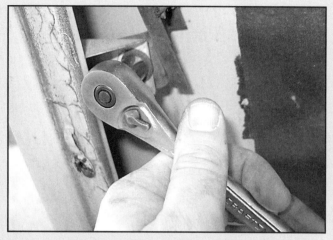

This 10mm bolt holds the lower window channel (and felt). At the top of this piece is a metal guard to protect the door mechanism from theft, so it won't just pull straight out but rather sideways.

Before yanking it out, this small clip must be bent up slightly, It keeps the lower channel close to the door so it won't bind the window on its way down or up.

The other half of the felt retaining clip snaps though holes in the aluminum stripping and into the car, holding the felt down, the aluminum tight and everything in place without glue.

We don't have a better picture of it, but the outer window scraper and aluminum trim are attached via a half-dozen rivets, of which you can see one here. Since we knew we weren't keeping the pieces and because they were so old, the rivets popped out with a little added pressure.

After we were done, here is the pile of discards. The only thing we'll keep is the vent window frame, glass and window glass. The rest can easily be replaced, so in the trash it goes.

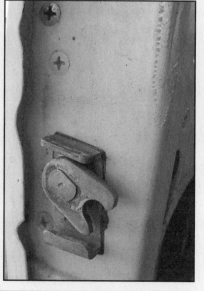

Doing its part to keep the door shut is this little gizmo, but hiding behind it is a complex system of levers and switches. It is attached to the door via three screws (two shown here above and below and one on the door side), the manual lock lever and the inside door latch. The screw above and to the left threads the door handle to the door.

REMOVING WINDOW GLASS AND DOOR LOCKS

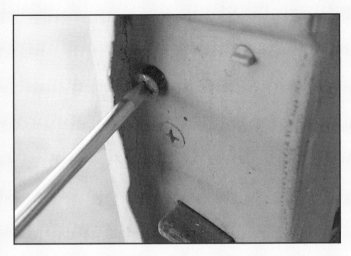

The first step is to remove the door handle, and you can do so by simply removing this screw.

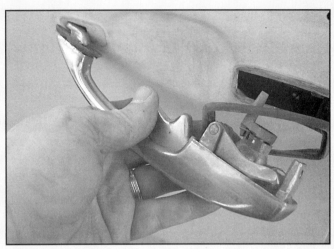

However, to get the handle free of the door, don't yank it off. Squeeze the lever, as if you were opening the door, slide the whole thing forward (toward the front of the car) and then pull it free.

Unscrew the door lock knobs, otherwise the door mechanism cannot be freed from the door.

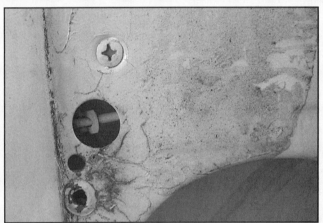

Here is the inside screw that holds down part of the lock mechanism. Remove this screw and the two on the jamb.

Here is an inside look at the mechanism...

...and finally out of the car. Easy. After some cleaning and greasing, it will be as good as new.

13 SANDING & PRIMING

Although we still have that fenderwell rust to contend with, there are some things we can do to help reduce our paint shop cost when we're ready for the shop.

New paint can be applied over old paint if the existing finish is sanded to a point where all oxidized paint material is removed and the surface is left flat, even and smooth. If bodywork is needed, the paint has to be stripped to bare metal so filler material will bond completely. Of course, each car is different. Some have been painted so many times that the buildup of layers is too thick to support another finish. For this car, since it has its original paint job, we decided that we would do a combination of sanding down some trouble areas but retaining the main body surface for the new coat. In other areas, for example, the hood and fenders, we've decided to acid dip them to bare metal.

For small dings and dents, body filler is applied to the area. The top layers are sanded with an 80- to 150-grit paper to smooth and flatten rough spots. Then 240 grit is used for additional smoothing. Use a block sander to keep the sanding uniform and consistent; vary your direction and let the sandpaper do the work. Every few moments, feel the surface to see your progress. Once the surface is free of ridges, use 320 grit to remove sand scratches and shallow imperfections. The 320 should be used to form a ring around the filler area, so you'll form a layered "valley" of filler, primer and paint. This is called "feathering."

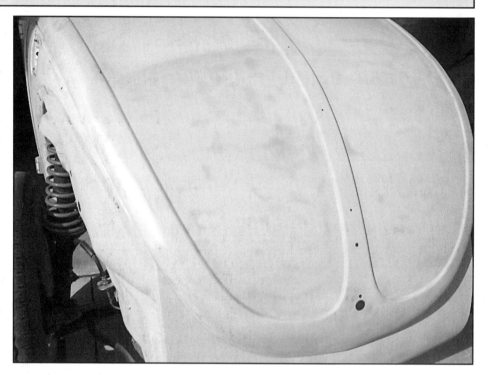

You can see the line where we stopped, about halfway up the hood. Below is primer, above, scuffed original paint. Pay close attention to the line between the two. It has to be perfectly smooth and flawless, otherwise any flaws will translate through to the new paint.

However you look at it, sanding is just as important a step as any other in the process, as every blemish will magnify if it isn't taken care of at this stage. If you don't scuff up the old surface before applying the new, more than likely the paint will begin to flake off, as it doesn't have a proper base to adhere to. Use 500- to 600-grit paper to scuff shiny paint finishes, because the overall purpose is to dull the finish so new layers of paint have something to grab onto. Sand in all directions and cover all surfaces to be painted, which in our case is the whole car.

At this point, the car and its parts need to be cleaned with a grease and wax remover. Every surface needs to be thoroughly cleaned. Once you get the car into the paint booth, this step should be done again (as well as the use of a tack cloth). Go over every inch that is to be painted—we can't stress this enough.

Masking

It would be easy to say, "mask off everything you don't want painted," but when doing the whole car, it is more involved than

PREPPING FOR PAINT

that. Consider the wheels, tires, the pan, the interior tunnel and floorboards, battery area, etc. Use masking paper that is specifically designed for painting, and whatever you do, don't use newspaper, as it is porous and paint will soak through. Don't forget to mask off the back of the dash, because you don't want any paint inside the trunk area. Now it's time to prime.

Epoxy primers and sealers do not have to be sanded, unless runs or imperfections show up on the surface. Then a fine-grit paper can be used to fix the blemish. Use the minimum recommended pressure and fan spray to gently cover the to-be-painted surface. Feather them into adjacent areas by slowly releasing the paint gun trigger toward the end of each pass.

Now that you have the major bodywork out of the way, you can focus your energy on getting the car painted.

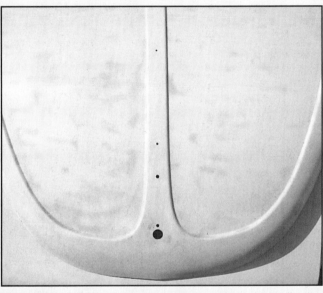
The idea here is to get it as smooth as possible. Since we had a lot of surface rust, we took this section down to bare metal and added several coats of primer, sanding between each one.

The areas under most of the windows are a great haven for rust to accumulate. Sometimes damaged window rubber allows water to sit on the sill, slowly eating away at the metal. Ours was in good shape.

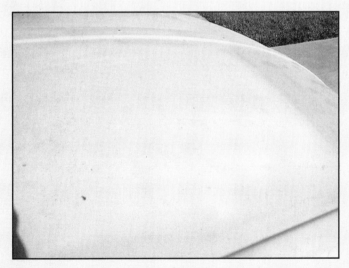
We were lucky that the hood was straight and smooth, even after all these years.

Here is a typical door, ready for sanding and priming. The best method here is to get a block sander and take your time. Use a contrasting-color primer as a guide coat to locate small imperfections as you sand.

THE VOLKSWAGEN SUPER BEETLE HANDBOOK

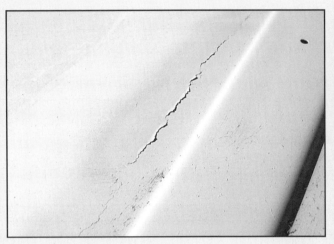

This is the ridge that forms when the original paint meets a bare-metal-primered surface. This blemish will ruin any paint job, no matter how expensive it is.

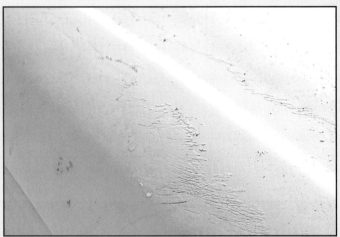

Unfortunately, sanding just one portion of the car (like this hood) leaves imperfections that will need to be tended to by using a finer-grit sandpaper.

This hood ridge is a prime place for rust to accumulate on the surface. For best results, make sure to remove all of the paint and primer to bare metal to check the integrity of the body panel.

If not fixed, bubbles will transfer through any new paint you apply.

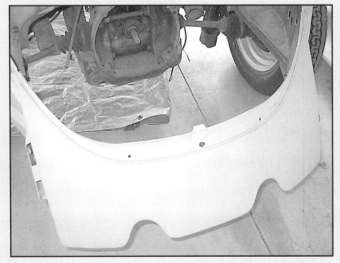

The finished results will be a perfectly smooth panel, such as this apron.

Surrounded by some rare metal, this is the Super before the body and pan parted ways. The pan came home with us while the body went up on a dolly.

SANDING & PRIMING

Our paint shop's prep crew began working on the Beetle once it was wheeled into the shop.

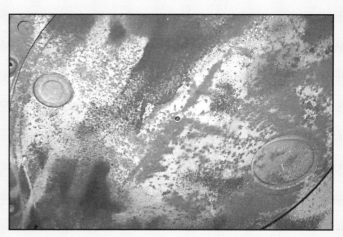

This is a shot underneath the trunk area of the body. As you can see, a lot of surface rust still needs to be removed. Even though nobody will see this aspect of the car ever (it's blocked by the frame head underneath), if we didn't remove this rust, it would soon eat its way into the trunk, painted or not.

A pile of fenders sits well worn in the corner awaiting their turn in the acid bath.

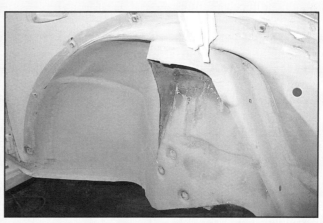

Because this car was hit pretty hard at some point, a layer of body filler was added on this side of the front clip. Most of it was removed, except for the area surrounding the welded-in fender nuts.

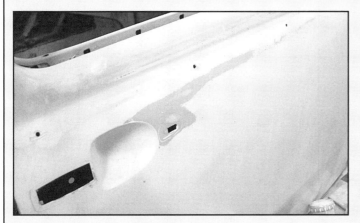

An important part of the painting process is to make the metal underneath as smooth as possible. When you add three or four layers of paint on top of a raised bump, you do nothing but make that blemish more noticeable. This can only be eradicated by sanding.

The next step for this car is a trip to the booth to get a couple of coats of primer, a few more hours of sanding and a few coats of top coat.

14 REMOVING THE BODY FROM THE PAN

In this chapter, we're going to show you how to remove the body from the pan, which will make it much easier to perform some of the bodywork and painting. The big thing to consider is that there are some differences from previous years—namely the struts—that must be contended with.

To do this properly, you'll need a few different-size wrenches, specifically 13mm 14mm, 15mm and 17mm. As well, jack stands and a floor jack will help keep the pan from buckling once the strut towers have been unbolted, and don't plan to start this without getting the help of a couple of friends to help with the actual separation.

If you are working with a complete car, there are a few things you'll have to do first (that we've already done in previous chapters) in order to remove the body from the pan. Specifically, you'll have to remove the seats (front and back), the fuel tank, the cable from the back of the speedometer and the battery. Though you really don't have to remove the engine, if you've got dual carbs, odds are good they're too big for the engine compartment to slip over them, so you'll have to remove them. If you do or not, don't forget to disconnect the wires to the starter and the accelerator and clutch cables. While you're back there, pull out the front engine tin so it won't hang up.

The body and pan are only held together at a couple of dozen points. A point of caution is to make sure you use plenty of WD-40 to help break up any rust or corrosion that may have built up. Since the last time the pan and body on this car saw separate ways was when they were first joined at the factory, the bolts were fairly stubborn (and we broke three).

With all of that peripheral equipment out and set aside, the remaining steps are easy, and we had a naked pan and an empty body in as little as a half-hour.

REMOVING THE BODY FROM THE PAN

There are a total of 18 bolts along the pan's side channels. A 13mm socket will make quick work of them, but if you break one, you'll have to drill it out and rethread a new hole. A missing bolt at this point might later cause squeaks or leaks.

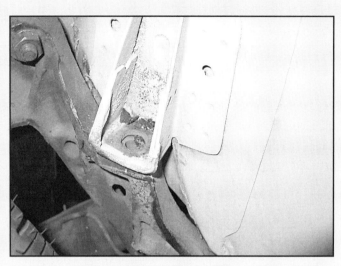

Remove the bolts on either side of the car where the body and the rear shock absorber mount meet. This is a 17mm bolt.

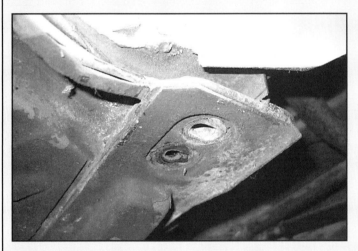

There are four 13mm bolts that hold the body to the bottom of the front crossmember.

Inside the trunk, underneath the two rubber grommets near the back of the spare tire well are these two bolts that connect the body to the front axle.

There are two of these bolts that hold the front the body to the frame head. Of course it'll have to be removed.

These three 14mm bolts hold the top of the strut tower to the body. There is no need to compress the springs as the towers can be pulled aside intact.

This is the universal joint for the steering column. There are two ways to do this: Disconnect it at this point only, or pull it out completely. If you leave it in like we did, you may need help feeding it through the hole as the body is lifted up.

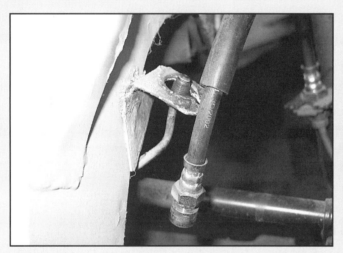

The passenger side brake line (where it clips to the body) can be unclipped and pulled aside, as there's enough slack in the line, but the driver's side line must be completely disconnected. You'll need a cup to catch the brake fluid.

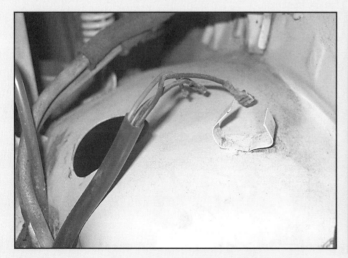

Regardless of what they are, make sure that all loose wires are properly tucked out of the way.

Underneath the car, disconnect the brake wires that attach to the master cylinder, as well as the brake lines that lead to the reservoir.

REMOVING THE BODY FROM THE PAN

The Super Beetle steering wheel and upper column connects to underneath the dashboard via two 13mm bolts.

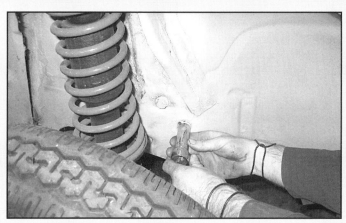

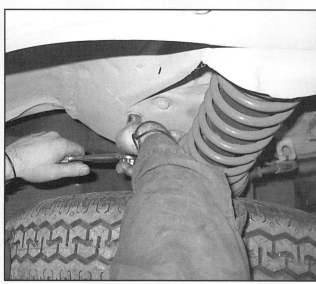

This is the only complication you'll need to tackle. The steering box and the idler arm bracket must both be unbolted from inside their respective front wheelwells. The steering box is especially tricky as it tends to hang up on the bodywork and must be coaxed from its spot while you lift the body. You may find it easier to separate the idler arm and the steering damper from the steering box and leave the box on the body.

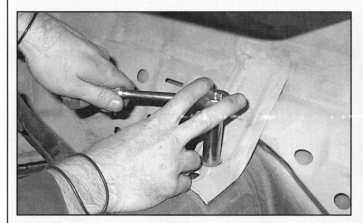

These two 14mm bolts help hold the front of the body to the tunnel. Notice that the body and pan is beginning to part at the front crossmember?

To keep the front of the pan from buckling because of the lack support by the front wheels and suspension struts, support the frame head with a jackstand.

As the body and pan were coming apart, notice the tight fit the steering box must navigate in order to stay with the suspension. Some persuasion was necessary until we gave up and merely disconnected it from the suspension.

After we felt that all was unbolted, we walked around the car and gave it a couple of test lifts, to make sure nothing was hanging up.

Before the body is finally free, here is one last look at the Super. The wheels are awkwardly pulled apart, but don't worry as nothing can get damaged as long as you keep unnecessary weight off of the wheels.

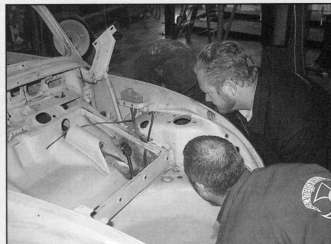

This is where your friends come in handy. Gather around and lift.

Since the next stop for this car is some extra bodywork, it will be rolled around on this dolly for the time being.

REPLACING RUSTED FENDERWELLS 15

You may remember in Chapter 7, when we tore out the package tray and replaced it with fresh metal. At the time, the fenderwells were hidden by a thick layer of undercoating, so evidence of the rust in that area wasn't clear until we had a chance to remove it all, which was in Chapter 10.

After pulling apart the body and pan (see Chapter 14), expert body man Jerry "Chopper" LeMieux assessed the situation and concluded that the job wouldn't be too difficult. If you plan on doing this at home, and there's no reason why you shouldn't, you'll need a reciprocating saw (or some such metal cutting device), a grinder, a hand drill and a wire wheel, along with a chisel, hammer and a wire brush. Most important, you'll need a welder, much like the Hobart Handler 135 model, specifically designed for 110 home and light shop use. We're using 35-gauge wire for good sheet metal. Speaking of sheet metal, all replacement panels for most all years Beetles can be found at BFY Obsolete Parts in Orange, California.

Since we're working with welding equipment, there are several safety measures you'll need to realize. First off, keep a fire extinguisher nearby. Second, protect your eyes with a a welder's hood and third, protect your hands with some welder's gloves.

Tucked in the corner of our body shop is our Super Project '71 awaiting new fenderwell pieces. As you can see, we've already done the passenger side, replacing the rusted-out tin of a gapping hole. The driver's side is still damaged.

THE VOLKSWAGEN SUPER BEETLE HANDBOOK

As you can see, the rust has actually turned the corner and is working on the area under the door hinges. Left unattended, it will eat into the heater channels and finally into the rest of the car.

First things first, explore the area. Use a screwdriver to poke and prod various areas to get the scope of the damage. How far up does the rust go?

Roughly mark off the area to be removed. Give yourself enough clean metal for the replacement piece to weld to. Our problem here was that if we went up any farther, we would begin to affect the double skin of the fenderwell where it meets with the interior skin.

To make the cuts as close to the marks as necessary, we're using a 0.035 inch cutting disk and air power. Though these disks don't last as long as thicker ones do, we can get a sharper, more exact cut.

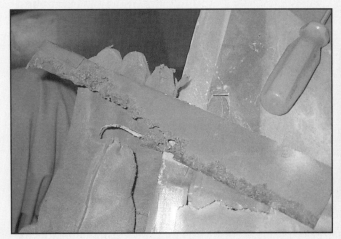

Once the piece has been removed, you can see the tendency of the rust to spread up the metal.

It would make sense to simply blow out the rust and debris with compressed air, but that would push the rust down through the body, on the outside of the heater channel. Instead, use a magnet to collect the scraps.

REPLACING RUSTED FENDERWELLS

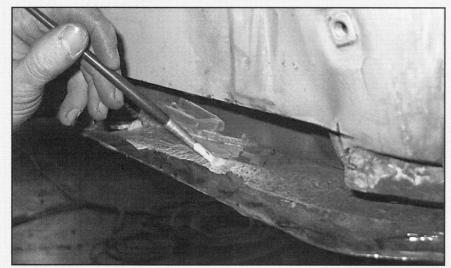

Included with the piece that we removed was the lip that curled down over another body piece, forming the lip that rests on the pan. You might be able to tell that this area is completely covered with surface rust. Unchecked, we'll be doing this process again in a few years when all that rust finally breaks through. For this reason we are covering all accessible areas that we can find with a rust-inhibiting sealer. A few well-known brands sell products like this—POR-15 and Bullfrog come to mind.

 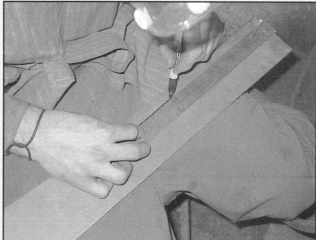

This is where the important part comes up: measuring. Measure the length of the opening and the height...don't forget to add approximately three-quarters of an inch to make up for the lip replacement.

 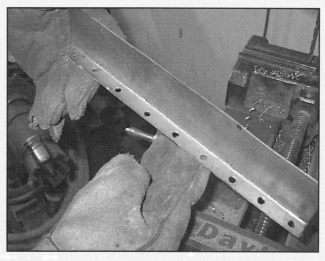

Once cut to fit, we bent over the lip and drilled several holes into it. Because we are using a welding technique called "drill and fill" these holes will be filled in with welding material and sanded smooth.

THE VOLKSWAGEN SUPER BEETLE HANDBOOK

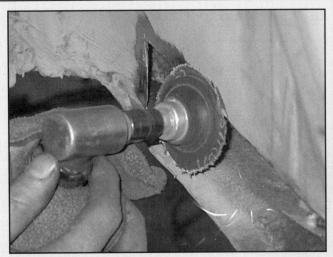

Before welding, wire brush (or coarsely sand) the surface clean of all primer, paint and undercoating. You cannot properly weld to material other than clean metal.

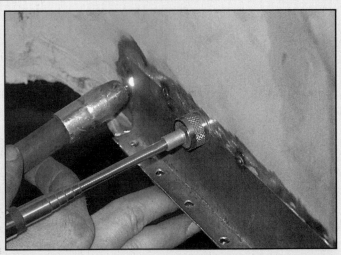

Line up the pieces perfectly and use a magnet to hold them securely while you lightly tack-weld the two together on the corners.

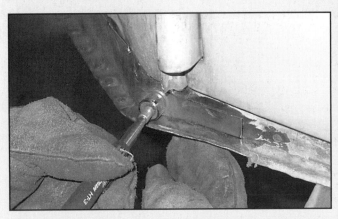

To make it around the corner, we did the same steps as before, except using a smaller piece. The smaller the piece, the less heat it takes to warp it, so go slowly. Spot-weld on one side, and then go to another side to let the first part cool. Use a misting of water to help with the cooling.

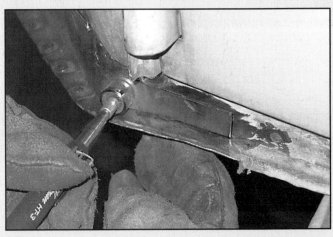

Once that is done, fill in the welds all along the perimeter of the replacement piece, making sure to fill in the drilled holes too. Sure, it looks ugly now, but the next step is to grind down all of the bumps and weld marks to a flat metal.

Once ground down, you can hardly tell work has even been done, especially in such a rough state as this.

A nice coat of primer and we're done, ready for final sanding and paint.

FINAL PAINT 16

Now that we were finally ready for the final paint job, we towed the Super Beetle down to our local paint shop, where we learned a few tips of the painting trade. We decided to paint the car in Kansas Beige.

If you're not sure what paint color you'd like, but know that you'll pick a stock color, check out www.wolfsburgwest.com, as they have an impressive collection of stock paint colors (matched to the stock interiors) categorized by model and year.

It is still possible to paint your car yourself if you have the time and the right equipment, as well as access to a paint booth. If you can't use a professional paint booth, it is possible to convert your garage, but you'll have to make it a "clean" room to make sure dust doesn't get into the paint surface. However, we can't possibly cover all you need to acquire and know to paint your car in one chapter, but we can show you the highlights of our Super Beetle's trip to the paint booth. I highly recommend that you purchase a good paint book before deciding to tackle the job yourself. HPBooks publishes three books on the subject: the classic *Paint & Body Handbook*, *Pro Paint and Body*, and the *Automotive Paint Handbook*. Information on ordering these can be found on page 170.

Here's the our old Super Beetle, sitting in the corner of our paint shop waiting for its in the booth

CARE OF YOUR NEW PAINT JOB

Once the car is painted, you shouldn't wax or polish it for at least 90 days. This will allow the finish to completely dry and cure. (When you are ready to wax, do not use waxes or polishes containing silicone or super polymers) It is recommended that you do not use a commercial car wash in the first 30 days (or at all, ever). Stiff brushes or sponges could mar the finish and damage the surface. Wash the vehicle by hand with cool clean water only. Be sure to use a soft cloth or sponge. Wash the vehicle in the shade only—never in the sun. Some other paint care tips:
• Do not dry wipe the vehicle... always use clean cool water or a quick detailer such as Meguiar's. Dry wiping can cause scratches.
• Extreme heat or cold should be avoided. Keep the vehicle parked in the shade whenever possible.
• Do not drive on gravel roads. Chipping the finish is easily done in the first 30 days.
• Do not park under trees which are known to drop sap or near factories with heavy smoke fallout. Sap and industrial fallout can mar or spot a new finish.
Keep in mind that trees attract birds. Bird droppings have a high acid content and will damage a new freshly painted surface.
• Do not spill gasoline, oil, antifreeze, transmission fluid or windshield solvent on the new finish. If you do...rinse it off immediately with water. Try not to wipe the area, pat dry.
• Do not scrape ice or snow from the surface. Your snow scraper will act very much like a paint scraper on a freshly painted surface.

THE VOLKSWAGEN SUPER BEETLE HANDBOOK

The first coat of primer is sprayed on.

Since they were stock from the factory, the doors were slightly wavy, so an extra coat of primer was added here and smoothed with varying grits of sandpaper.

After the second coat of primer, the whole body is wet sanded with 400-grit paper to a dull, but smooth shine.

Finally, after the top coat of Kansas Beige is applied, the body is put back on the pan and it is ready for reassembly. The body was color sanded, but not yet buffed and polished to a high-quality shine.

ASSEMBLING THE LONG BLOCK 17

Building your own engine from a pile of parts isn't something to be taken lightly. For a novice engine builder, the many boxes of parts that will arrive at your doorstep can be a bit overwhelming. But there is an easy alternative: Let someone else assemble the long block while you assemble and install the ancillary components, like the carburetor, alternator, fan housing and engine tin. In the long run, you'll save money over buying a turnkey engine, and you'll learn much more about what powers your Volkswagen.

To do this, we contacted The Real Source in Effingham, IL, and ordered their "Factory Fresh" 1600cc long block, straight from VW's original equipment supplier. This long block comes complete from valve cover to valve cover. Best of all, the complicated parts of the engine have been built for you. Better than that is that the parts are all brand new, not rebuilt. All you have to do is add the fuel supply, carburetor, intake, cooling and exhaust.

Since this could take a while to put together, it is best to dedicate a section of your workshop and purchase an engine stand. The Real Source can set you up with one, but ours came from EMPI. The tools you need include screwdrivers and a handful of the usual metric sockets to do the job from start to finish. Since our engine arrives from the factory, it was coated with Cosmoline, a sticky petrolatum

We have a lot to do in order to make this engine purr like Porsche intended. In this photo, the fuel pump, generator stand, distributor and oil cooler, have already been installed, but we'll cover those steps in the photos that follow.

used to protect the parts from rusting during shipping.

The first order of business is to get everything straightened out. Find out what you have and what you might still need. We discovered that none of the parts came with the required hardware, so we sent away to Totally Stainless in Gettysburg, PA, for their engine and exhaust stainless steel fastener kits. As well, we wanted to make this engine look as good as it will run, so we packed up our 25-piece fan shroud kit and sent them off to our local powder coater for some protection and we shipped out the exhaust pieces, heater boxes, intake pipes and even the alternator stand to get them ceramic coated.

One thing you'll notice is the beautiful shine some of our new parts have, namely the exhaust system, heat exchangers, the intake manifolds and preheater pipe (even the generator stand has a sheen to it). On most engines these parts are, if not rusted by time and moisture, discolored from heat. But we won't have that problem on this engine. For starters, our parts are brand new, but we took it a step further and dropped the heater boxes, exhaust, manifolds and stand to A-1 Muffler in Santa Ana, who, in partnership with Engineer Application in Brandon, Calif., were able to not only protect the parts from the elements but give them a lasting shine that will really improve the overall look of our engine compartment.

In contrast, we had all of the tin properly powder coated by our local coating shop for equal protection from the elements. The black will go well with the manifolds and exhaust.

This is our fan shroud kit after getting a nice treatment from our local powder coater. Powder coating not only looks nice but gives the metal a protective layer of "paint," keeping the tin from rusting and discoloring.

This is the collection of most of the major parts we'll be installing on this long block. It may seem like a lot, but most of it is fairly straightforward. Make sure you keep everything organized so you can find parts quickly as you need them.

All of the holes on the long block have been covered to prevent debris or dust from entering the engine and cause some major damage. Good advice is to keep these holes covered until you need them. Nothing is more defeating than dropping a nut down the fuel pump hole.

There are three things you must align when installing a distributor: the driveshaft slots, the line for number one piston stamped on the body of the distributor and the position of the rotor. To do this, start by rotating the crank pulley so that number one piston is at top-dead-center. Slip the rubber O-ring on the distributor's body and place it in the hole so that the tangs on the housing mate with the slots on the driveshaft deep in the hole.

After the tangs match (they can only go one way), rotate the body so that the number one piston line is matched up with the rotor. Tighten the clamp (which doesn't come with the distributor by the way, so you'll have to find one) and you're set.

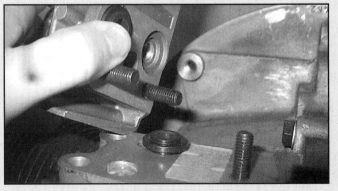

VW used four types of grommets for their oil coolers: early Beetles and Buses; Type III to 1969; 1970 and later for all types, especially doghouse coolers for 1970–'79 Beetles and 1970–'71 Buses; and tapered grommets used to adapt new 10mm coolers with old 8mm cases (changed in 1970). These are installed without using any kind of sealer so as not to clog the oil cooler.

ASSEMBLING THE LONG BLOCK

Three 13mm nuts tighten down the cooler mount to the case. It is a good idea to feel the mating surfaces for bumps and imperfections, as you'll want a smooth connection to prevent leaks.

It is tricky to keep the grommets in place while you attach the cooler to the mount, but with a couple of tries, you can persuade them to stay as you put the two bolts through their holes.

Use an 11mm socket and don't forget the lock washers, which are actually supplied by the manufacturer.

Next we turn to the pedestal. Use a thin coat of a sealer (in our case Permatex) before placing down the louvered metal filter (which prevents blowback). The louvers should be down and facing the flywheel like shown.

13mm nuts, washers and socket complete the installation.

Apply sealer to the case, then set down the lower gasket.

THE VOLKSWAGEN SUPER BEETLE HANDBOOK

Add sealer to the underside of the insulator block and place it on top of the gasket.

Apply some grease to the fuel pump pushrod and then slide it into the insulator block with the pointed end down.

With some sealer on the bottom of the top gasket, put it into position and then add some sealer to the top of that gasket.

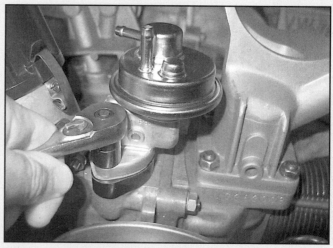

Finally, a 13mm socket is used to tighten the fuel pump to the case. Since it can only go in one direction, you can't force it the other way.

The next step is to install each of the two cylinder cover plates. There is a right and a left, so make sure that you get the correct sides. As well, you may have to do some modification so that the manifolds will clear the lip of the tin.

At the rear of the engine, place the flange gaskets (dry) and bolt on the heat exchangers. There is a left and a right for these as well, so make sure to note which is which—they're actually labeled with an "L" and an "R" on the underside.

ASSEMBLING THE LONG BLOCK

Moving to the front of the engine, place the metal gaskets (also dry) on the exhaust ports and fit the exhaust system. Some stretching was necessary to get the ports to line up properly to ensure a tight fit. Make sure to include the gaskets for the heating junction boxes on both sides.

Next, attach the intake manifold (photo at right) to the intake ports with 13mm nuts and washers. You may have to modify the cylinder cover tin to make the connection solid. In this case, we started on the right side, but it doesn't matter. Since the manifold has to slip under the generator stand, it is easier to do it piece by piece. Clamp it together with the polyurethane clamps.

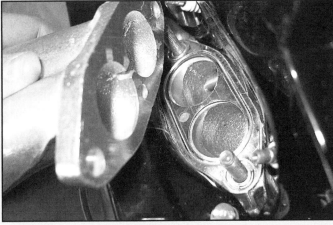

The intake gasket is placed dry and the manifold on top with 13mm nuts. The opposite side goes on last because it can be fitted to the cross piece.

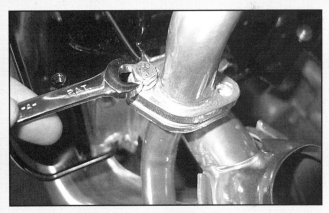

Don't forget gaskets on both sides of the exhaust flange. The flange is bolted to the exhaust system via four 10mm bolts and washers. You may have to loosen almost everything (the heater boxes, exhaust and manifold) to make sure everything can be adjusted to fit.

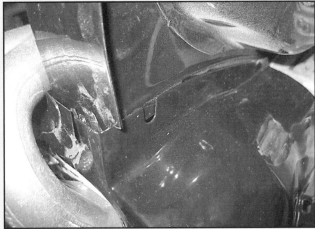

For the sake of fitting, we placed the doghouse fan shroud on top of the cylinder cover tin to check for location. There are two engine tin screws on either side that hold the shroud in place. This is temporary and is easier to do now without all the weight from the generator. You may need a flat-head screwdriver to pry open the cylinder tin slightly to accommodate the shroud.

THE VOLKSWAGEN SUPER BEETLE HANDBOOK

Over on the bench we've laid out the parts needed to assemble the generator to the fan. Missing in the shot are three spacer washers.

This little Woodruff Key is quite important to the whole process. If you assemble it without this key, odds are good you'll start spinning the fan hub (which might cause the fan to malfunction) and ultimately damage the engine.

Place the two cover plates onto the generator and bolt them down. Use Loctite and lock washers, as these bolts aren't something you want to lose inside the fan shroud.

The Woodruff key slips into the slot on the generator shaft. Make sure it is properly seated. Add a spacer before the thrust bearing.

The thrust washer and fan hub are place on the shaft next. While the hub is slipped on, make sure it is lined up with the Woodruff key and that they key isn't pushed out of alignment.

Use the three spacer washers to achieve a 2mm distance between the fan and the fan cover. If you use one spacer between the hub and the thrust washer, the other two should be placed between the lock washer and the fan. The nut should be torqued to 40–47 ft-lbs.

ASSEMBLING THE LONG BLOCK

While still at the bench, slip the fan into the shroud and attach the cover with four tin screws and washers.

Here is our progress so far. The fan shroud should fit tightly over the cylinder cover tin. Screw it down on both ends of the shroud.

The generator strap easily slips around the stand and is tightened with a 13mm wrench.

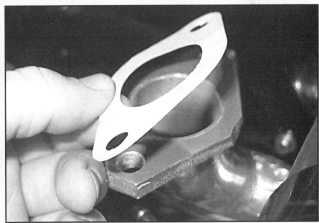

Next up is the carburetor. The paper gasket goes on first.

It is a bit difficult to reach around to the back to thread the rear 13mm nut and washer, but not impossible.

Place the inner half of the pulley on the generator. When working with the fan belt, you'll want to have relatively clean hands, as any oil on the belt can cause it to slip under stress.

Add as many spacers as you need to obtain the proper belt tension between the two pulley halves. The belt, when pressed down between the pulleys should yield approximately 15mm (which is 0.6 inches). Use a screwdriver in the slot on the inner pulley half to hold the pulley while you tighten the nut.

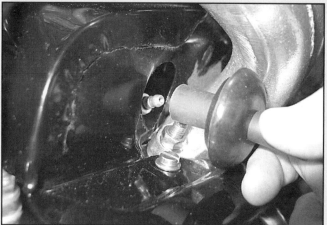 

Attach the spark plug wires to their respective spark plugs. In your kit, you're given two long wires and two short wires. Obviously the long ones are for the opposite pistons and the short ones are for the pistons nearest the distributor.

 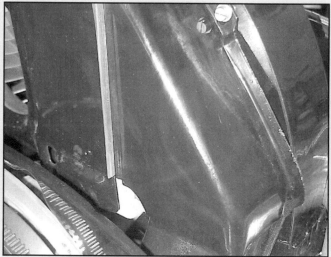

The coil lead arrives "unassembled," meaning that the end connection needs to be attached. This gives you the freedom to mount the coil in any place you want on the engine, either in plain sight or hidden from view. We decided on the stock-type location. If your shroud doesn't have holes, you'll have to drill two to mount the coil.

Three more pieces are installed. The snorkel blows hot air off of the cooler and out of the engine compartment, while the rear engine seal tin keeps the compartment closed tight.

ASSEMBLING THE LONG BLOCK

There you have it. Simple and easy, the ancillary equipment is on our engine, but we're far from ready to stick it in our Beetle and fire this thing up.

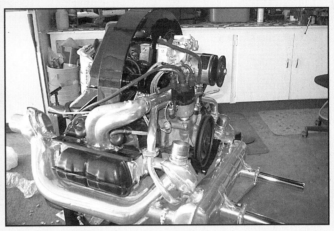

At this point, we need some professional help because we ran into a few problems. For instance, we were sent the wrong rear tin and had to make do with one from another kit...but the catch was that it wasn't cut for pre-heater pipes. So, we took the engine to Clyde Berg and had him finish things up.

First, he leveled the fan and generator by adding this shim. The result is a smooth spinning fan that can't rub on the shroud.

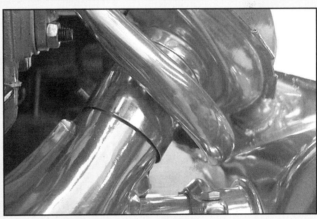

Brackets were added to the heater boxes where they connect with the muffler. The exhaust system has to be airtight otherwise you lose performance and you could introduce unwanted gases into the cabin.

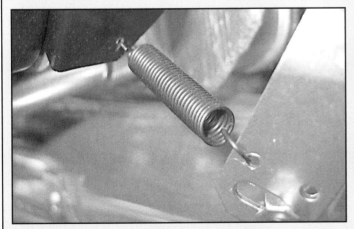

The bracket system for the heater boxes as they attach to the tubes that direct hot air into the car. Since we'll be keeping the heater system on the car this engine is intended for, it was critical to make this work. Leave it to the Germans to design a needlessly complicated system that only opens a butterfly valve.

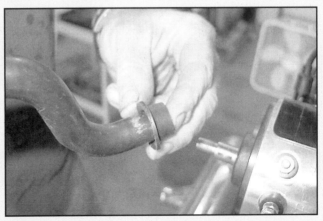

We had to buy a couple of missing items from Wolfsburg West, one of which was this rubber grommet for the oil breather tube. The tube itself is a used one from Topline's cache of parts, because the shiny chrome one we wanted to use wouldn't fit around a small (useless) knob on the generator stand.

THE VOLKSWAGEN SUPER BEETLE HANDBOOK

The oil breather tube attaches to the generator stand with this nut that fits inside of the tube. There's a special tool to tighten this nut, but needlenose pliers work just as well. Make sure it is tight; otherwise it will leak.

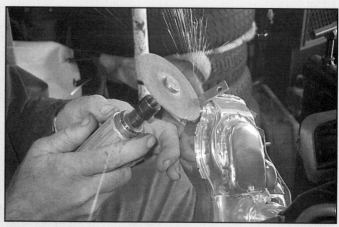

These heater box tabs are used for Buses (this is the only difference) and are not needed on Beetles. Since they will hit the underside of the car once the engine is in place, they had to be removed with a grinder.

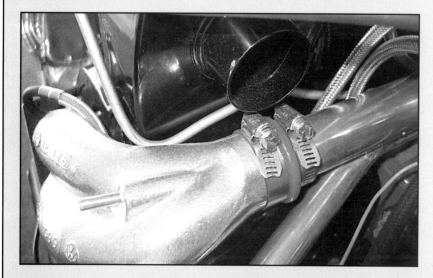

Another part we needed was this metal gas line that runs around the fan shroud. Again, we picked this up at Wolfsburg West. It slips in under the manifolds and around the shroud.

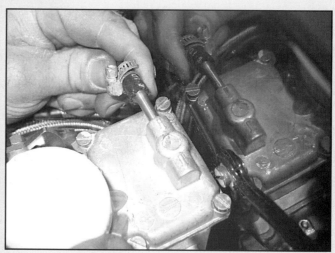

The inlet fuel line attaches here to the horizontal post on the fuel pump, while the outlet tube attaches to the upward-pointed post. It goes up the carburetor and into the float bowl at the back of the carb.

ASSEMBLING THE LONG BLOCK

Instead of engine tin screws for the sides of the fan shroud, Clyde only uses 6mm bolts. That way, when everything is in the engine compartment, you can reach it with a wrench rather than a screwdriver if the shroud has to be removed.

Here was our first big problem. The rear tin that came with the kit didn't fit to the tin that all new engines come standard with. So, this tin is old tin that has space cut for preheater tubes...but obviously doesn't fit (above or below). And if it doesn't fit, it doesn't do its job.

To solve this problem we used a piece of tin from another engine that we knew would not only fit around the existing tin but have holes for air inlets to the shroud.

The only problem was that it didn't have cutouts for the preheater tubes (where they connect to the exhaust). No problem a grinder couldn't solve.

Since our 009 distributor doesn't need to be equipped with a vacuum advance, there's no need to hook up the carburetor to anything. So, a small piece of vacuum tube and a screw will block off the vacuum port but keep it functional if we need to use it in the future.

 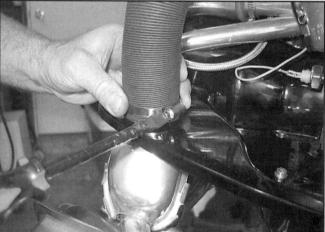

One of the last things to do before we shoehorn this engine into the car is to cut and install the heater hose that goes from the fan housing to the heater boxes. Each one is cut to length and attached with hose clamps.

REPLACING THE IRS TRANNY 18

The AH-case IRS transmission that came stock in this car comes with a final drive of 4.125 (as do all Supers in 1971 and '72, but thereafter it was decreased to 3.875 for the AT cases). First gear ratio runs at 3.80; second at 2.06; third 1.26; fourth 0.88; and reverse runs at a ratio of 3.61:1.

REMOVAL

Removing the transmission is a fairly easy operation. The transmission itself (without the axles) only weighs a few dozen pounds, and with our method of removing it, you won't have to worry about it falling on your foot (or head) while you're unbolting it.

The required tools are an assortment of metric wrenches, a Phillips screwdriver, jack and jackstands, and one specialty tool: a 12-point driver to remove the socket head screws on the CV boots, which can be found at a wide variety of VW shops.

Once all of the tools are assembled and you're ready to go, you should start by disconnecting the battery. Since we'll be removing live wires from the starter, you'll not want to short anything out or fry any wires. At the front of the car, block the tires and pull up on the parking brake. To make this procedure easier to demonstrate, we've decided to do this while the body and pan were separated. While it is easier to work on without the body in the way, you can do this easily with the body still on the pan.

Here it is as it now sits, an original transmission that has been part of the car since the day the car was born. Covered in grease and dirt, there's probably really nothing wrong with the internals of the trans, but with the body gone and the engine missing, it is a great time to make a switch. Plus, we'll need to replace the CV joints and boots along with the shift rod bushing and a couple of other things anyway, so to take out the tranny only takes a few extra steps. With a transmission as steadfast as this one, it is a good idea to invest in a can of penetrating oil and we only use WD-40, of course.

INSTALLATION

Installing a transmission is not quite as simple as reversing the removal steps. Though you are replacing the same bolts in the same holes and the end result is exactly what you started, the order is different and there are a few changes that need to take place.

We contacted Mark Wolter at Strictly Föreign in Grant's Pass, Oregon, for one of his "freeway flyer" transmissions. Since the tranny is headed for the back end of our Super Project '71, we wanted one that would enable us to cruise at traffic speeds without too much stress on our engine and the "freeway flyer" fits the bill perfectly. Inside the case sits a 4:12 ratio ring and pinion, with helps with the new gearing ratios: 3.80, first; 2.06, second; 1.26, third and 0.82, fourth.

In addition to the new gearbox, we are also adding EMPIs complete axle packages to both sides of the tranny. The kit includes axles, CV boots and grease. While we had everything apart, we pulled out the shift rod tube to have a look at the bushing, which needed replacing.

THE VOLKSWAGEN SUPER BEETLE HANDBOOK

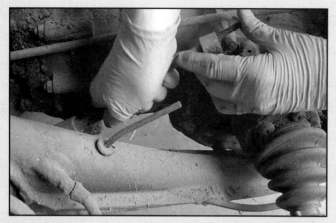

The first step is to remove the clutch cable, and to do that, start by slacking the cable by pushing back on the clutch operating lever (as if you were pressing down on the clutch) and place a socket underneath the operating shaft. Once the tension is released on the cable, use vice grips to clamp down the cable and then twist off the wing nut. This is slightly difficult, but you'll need to pull down on the cable housing so it is free of not only the operating lever but another bracket further forward on the case.

Moving to the starter, use a 13mm socket to remove the wire from the battery to the starter solenoid. This is the live wire and you'll soon know it if you didn't disconnect the other end from the battery.

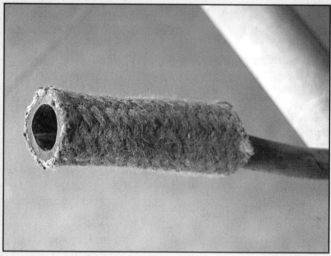

A small trick to keep the terminal end of the wire clean and in good shape is to cover it with a piece of fuel hose.

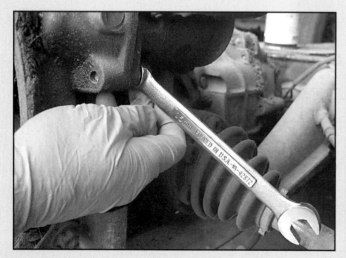

The bottom bolt that holds the starter to the transmission is of the 15mm variety and is easily removed. Since we'll be replacing this starter, there's no need to keep it any longer than its trip to the trash can.

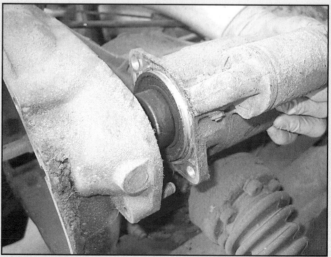

Once the bolts are removed, pulling out the starter is as easy as breaking the caked-on dirt that has sealed it in place these so many years.

REPLACING THE IRS TRANNY

Before you start to pull apart the socket head screws that hold the CV joints to the trans case, make sure you clean out the splines inside the screws so the 12-point driver mates fully with the inside of the screw. If you don't, you could strip out the screw and then you'll have to cut it off and drill it out.

First, unscrew the top three on each side first.

Then jack up the back end of the car by placing the jack under the center tunnel. Don't forget that safety is first; use jackstands and block the tires.

After putting the transmission in gear to lock the wheels, unscrew the remaining three screws on either side

The stock bolts that hold the nosecone inside the front trans mounts are supposed to be a strange 15mm head/10mm hybrid bolt. Usually, after a rebuild/R&R, they get lost and their 17mm counterpart are used instead. The problem with the 17mm is that it is difficult to get a wrench into this small area. If that is your case, you'll need to support the transmission and remove the mounting nuts instead.

To get to the shift rod coupler, remove the inspection plate under the rear seat with a Phillips head screwdriver. This is the only place to access the tunnel, so if you butterfinger something down there, you'd better be able to reach it or live with it in there forever.

103

THE VOLKSWAGEN SUPER BEETLE HANDBOOK

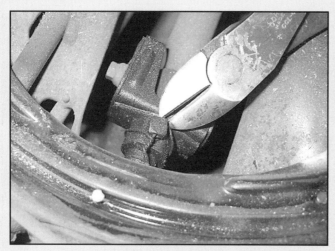

Once inside, snip the safety wire that surrounds the bolt at the point closest to the transmission.

With an 8mm wrench, remove the square bolt that holds the coupler to the shaft.

Then push the gear shifter into second gear. This should be enough to pull the shift rod away from the inner shift lever.

Once the rod is pulled back, it should let go of the coupler.

With the front of the transmission detached, it is still being held to the frame via two different ways, the trans bushings and the two 27mm cradle bolts. Because of this, there are two ways to go. We went the bushing route for two reasons: One, we wanted to use the cradle to continue holding up the transmission; and two, most people don't have a 27mm socket.

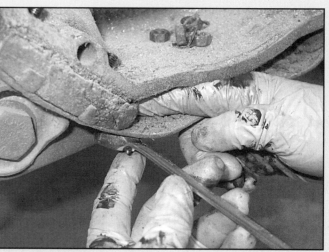

The trusty 13mm wrench is at it again. This time removing the four bolts that hold the trans bushing to the transmission...

REPLACING THE IRS TRANNY

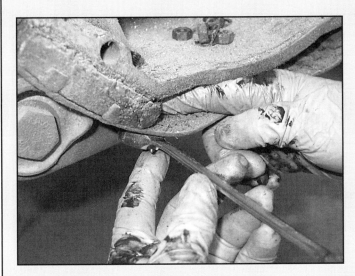

...and the cradle to the transmission bushings.

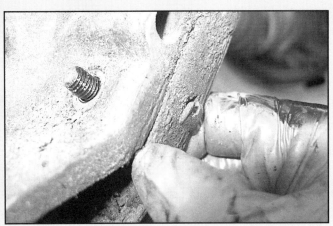

This is a step we'd eventually have to take because of the deplorable condition of our bushings. Notice that the metal casing of the bushing has separated from the rubber. After all of this work, in our opinion, it is a good idea to replace the bushings regardless of their condition.

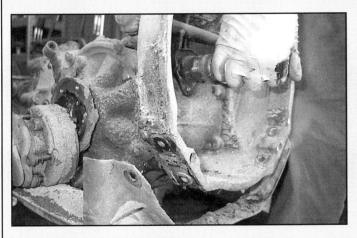

The transmission is lifted off of the cradle and out of the front mount, and the bushings fall to the floor.

The axles and the CV boots will hang off to the side for now. We'll tackle them in the next section of this chapter. Since the axles are incredibly dirty, consider wrapping them in plastic bags until they are replaced.

Fortunately, most all of the rubber grommets are available from your local VW shop, including this housing bushing.

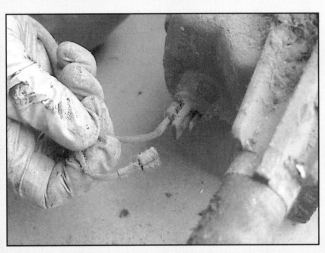

This is the wire for the reverse lights. We'll touch on this in Chapter 21 later in the book.

THE VOLKSWAGEN SUPER BEETLE HANDBOOK

After only an hour of work, the transmission is out and ready for rebuilding.

Since it's out of the way, remove the front mount bushing by removing these 17mm bolts.

The pan has the original axles and CV boots that had never been removed. Since the body is off the pan, it will be an easier job than if you had to work around it.

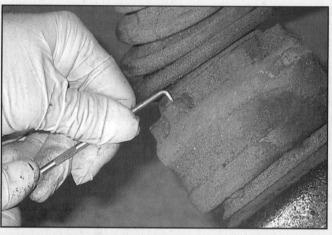

We used this small dental-like tool to remove the dirt from around the bolt heads. That way our 12-point socket will have a better hold.

The other half of the axles come off with the same method we used in the first part of this chapter, but make sure you have the proper 12-point driver to remove the bolts. For every pair of bolts, there is a retaining clip that helps distribute the pressure all the way around the axle.

The new axle assemblies are strapped down so the CV joints don't unhinge themselves and fall out. They must be packed correctly, otherwise they will fail.

REPLACING THE IRS TRANNY

When you are putting the axle grease in the CV boots, make sure to fill it up as much as possible. Fill every nook and cranny until it squishes out of the seams.

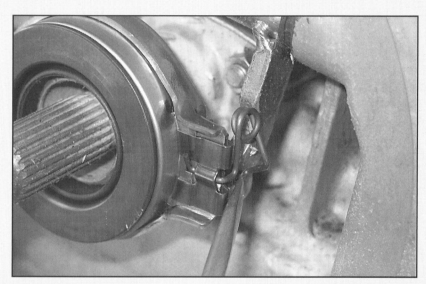

Moving onto the transmission, we installed the throwout bearing by slipping the retaining spring over the operating shaft.

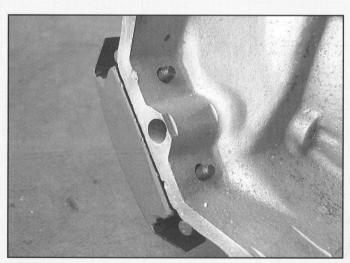
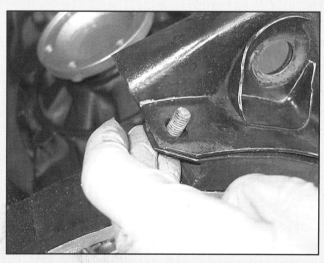

Because of the angle of the mounts against the transmission, it is easier to attach them to the transmission and the cradle bushings than it would be the other way around. Leave them loose until the end.

THE VOLKSWAGEN SUPER BEETLE HANDBOOK

The bushing for the clutch cable and the accelerator cable simply slips over their individual protective sheaths.

At the other end, bolt the nose cone bushing to the chassis (via a 17mm wrench) and make sure it is right-side up (the U-shaped ridge needs to point up as shown).

The nose cone bushing helps with vibration and smoothness in shifting.

Align the cradle mount strap with the 27mm bolts and tighten, then go around and tighten all of the bushing bolts (there should be eight altogether).

REPLACING THE IRS TRANNY

At the other end are the odd-sized stock nosecone-to-transmission bolts, which are a 15mm head/10mm thread hybrid bolt. Hopefully, you remembered where you put them when you removed them—they aren't easy to replace.

With a jack and jackstands, lift up the rear of the car so the rear tires are slightly off the ground. Rebolt the axle assemblies literally the same way you removed them, three at a time on each side and then rotate the wheels for the bottom three.

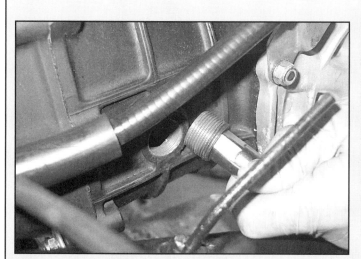

The final thing to do is fill the case with oil and you're ready to go.

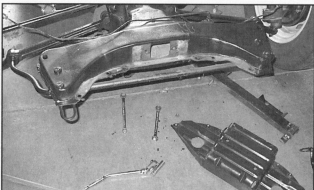

Since we had the car apart and it wasn't that difficult to do this extra step, we wanted to show you how to replace the shift rod bushing, a small grommet that usually fails. The first step is to expose the shift rod tube hole by pulling off the front frame piece. Like all Volkswagens, there is only one way to install and remove a shift rod tube and that is through the front. There are eight bolts that hold this sheet metal to the frame; don't forget the two horizontal bolts.

THE VOLKSWAGEN SUPER BEETLE HANDBOOK

After you have unhooked the rod from the shift rod coupler (under the plate under the backseat), which actually we hadn't hooked up again after replacing the tranny, use pliers to shimmy the shaft toward the front of the car.

The goal is to pull the rod out of the hole. This may take two people to accomplish, one to push and one to pull.

The shift rod bushing in question is this broken plastic ring that fell off the rod as soon as we touched it.

Also while we are reaching down into the shifter hole, we inspected the bracket that holds the shift rod up through the hole. It looks a lot like this, and if yours is broken, missing or loose, now is the time to fix it. Unfortunately, that would require some welding and a couple of extra hours. Ours was perfect.

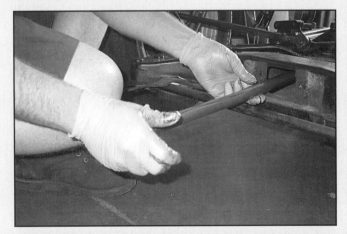

The next step is to polish the rod to a nice shine, basically to clean off any rust, gummed up grease or dirt. After a nice coat of white grease, slide the shaft back into the hole and push it back as far as you can.

Before the rod reaches the shifter hole, place the new shift rod bushing over the bracket. Then line up the rod with the bracket and bushing and give it a hard shove so the bushing slides over the rod smoothly. Finally, attach the shift rod coupler to the shift rod and the transmission shaft. Replace the sheet metal on the frame head and you're done.

INSTALLING THE HEADLINER 19

There's a myth among car enthusiasts that the headliner of any car is very difficult to install. This is true to a certain extent, but it is not impossible for an ambitious do-it-yourselfer to do. There are really no special tools needed or any special skills needed—although experience certainly is on the side of Octavio Gutierrez, owner of Octavio's Kustom 1 Upholstery shop in Orange, Calif., a secret source of not only great custom and classic interiors but for overall restorations.

Tools necessary for this job are a pair of sharp scissors, a heat gun (a strong hair dryer will help), razor blade and upholstery adhesive. Start by laying out the large headliner in the sun for a couple of hours, as this is one job that is better done on a hot day to keep the vinyl malleable. There are a multitude of options for a headliner, from leather and vinyl to wool and cloth, but we chose a stock standby TMI headliner from BFY Obsolete Parts. Another step you should consider is to soundproof the top. Two sheets of sound deadening material such as Dynamat will do the trick in keeping out a lot of Mother Nature's noise. The headliner has a film that pulls away to expose the adhesive side.

This is Octavio's place. Though small, it is well-staffed with a talented crew that works together to ensure efficiency on each job they have scheduled.

111

THE VOLKSWAGEN SUPER BEETLE HANDBOOK

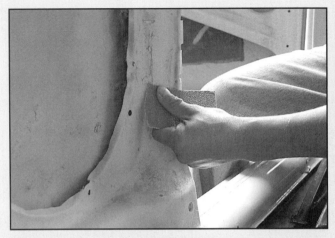

Since the Super just came back from the paint shop, the overspray must be sanded down to bare metal, enough to form a roughed up surface that the glue will stick to. This is done with 36-grit sandpaper.

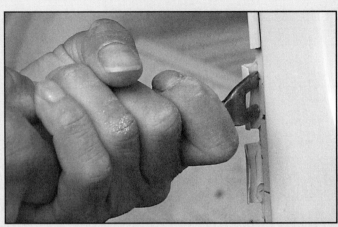

The retaining teeth on the door pillars are bent out and straightened. This can be done with a pair of pliers. You're going to lose some paint in this step, but don't worry, it will be hidden by the vinyl A-pillar pieces.

After all of the paint chips and dust have settled, vacuum out the debris and make the surfaces as clean as you can. Glue will only stick to clean surfaces.

Octavio measures for the sound deadening material. We ended up using two pieces, each approximately 32 inches wide and totaling almost 60 inches long, but don't take our word for it, measure it yourself.

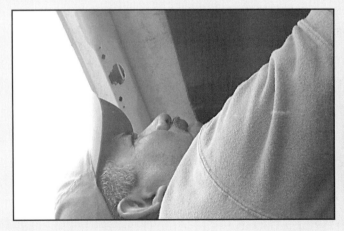

With the heat gun, the material is heated up, causing the glue to run slightly before it is put into place. After a few seconds of holding it there, it will stay nicely.

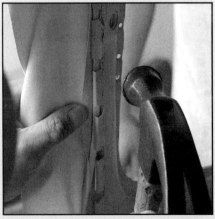

Starting with the A-pillar piece, feed the rubber piping into the toothed clamps and hammer them down tightly. Start at the top of the door and go down, removing excess material at the bottom.

INSTALLING THE HEADLINER

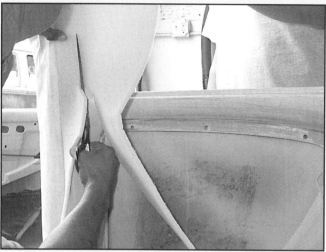

Add a coat of glue to the back side of the vinyl and the door pillar. Then cut a piece of foam padding with enough clearance to wrap completely around the pillar, approximately three inches.

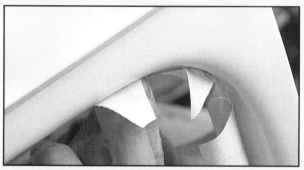

With another coat of glue, tightly pull the vinyl around the foam and to the door pillar. The glue should be tacky enough to stick well after only a few seconds. Start in the middle and go up then down, making sure the foam stays smooth underneath. If there are any wrinkles, smooth them out with the heat gun and by stretching the vinyl farther. At the corners you must cut slits in the material (shown above) to not only relieve stress at these points but to allow the flat vinyl to smoothly make it around the corner.

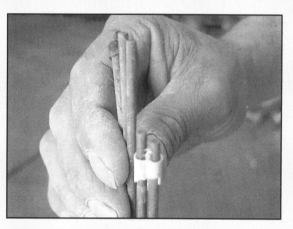

You'll notice that with the headliner rods, there are two shorter than the others. These two go in the front of the car, nearest to the windshield, while the other three (four if your headliner is original—aftermarket headliners only used three of the rods for some reason) go toward the back.

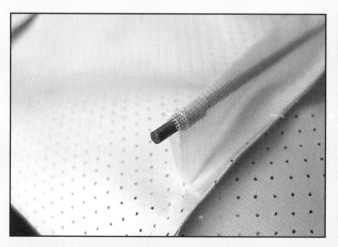
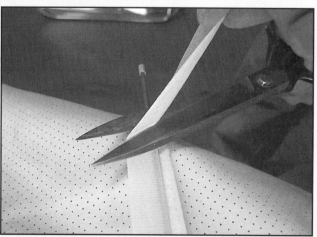

The rods are fed into the main headliner piece and the ends of the channels are cut back a couple of inches to give the rods more space to bend.

113

THE VOLKSWAGEN SUPER BEETLE HANDBOOK

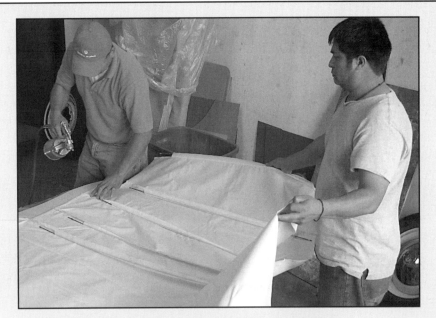

A coat of glue is added to the entire perimeter of the headliner as well as any point on the car where it is needed. Assistance is needed to keep the vinyl from sticking to itself once the glue gets tacky.

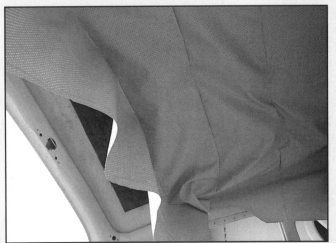
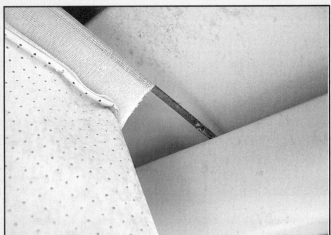

Each of the five rods are placed into the channel above the doors and windows, the shorter ones farther forward. Don't worry too much about spacing at this point, because when you stretch out the headliner, the rods will fall into place on their own.

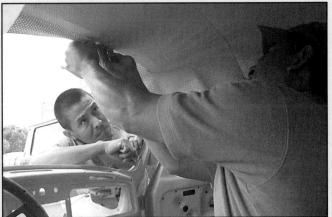

Starting in the middle of the front (windshield side), Octavio pulls the headliner tight against the window frame and presses the glue down. This is one of the most important steps, because if it is off center, the whole headliner will fit crookedly.

In the back, pull the main piece as tight as possible. This is where the rods will fall into place. Then move to the front, as shown here, and pull the material above the windshield as tightly as at the rear.

INSTALLING THE HEADLINER

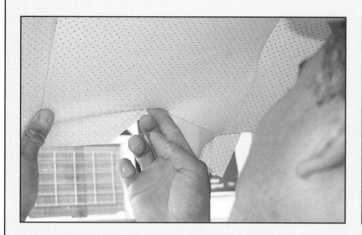

Another important step to get perfectly correct is the position of the center rod, as it should be approximately two to four inches to the rear of the center B-pillar. Make sure it is even on each side. Note the fancy fold on the B-pillar. This is done by simply tucking in the headliner with a slight downward angle.

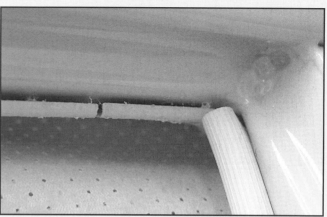

These are the exposed clamps for the sides of the headliner. They can be sharp, so be careful. Getting blood on your new headliner isn't the best way to complete the job.

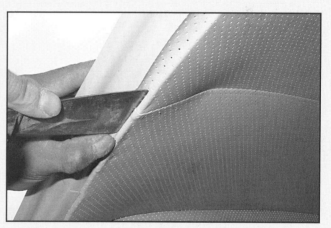

From that center point pull and feed the material into the toothed clamps above the door. Cut the headliner so there's about an inch of overhang so there's enough to grab hold inside the groove.

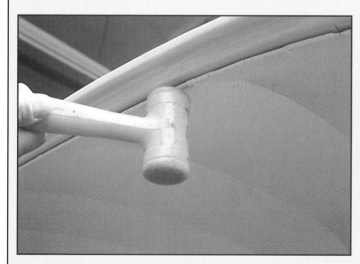

With a plastic or rubber mallet (anything that won't leave a mark), tap down the clamps.

Working back toward the front pillars, the material is stretched tight and a similar fold is executed as on the B-pillar. Just make sure that they are at the same level on both sides.

THE VOLKSWAGEN SUPER BEETLE HANDBOOK

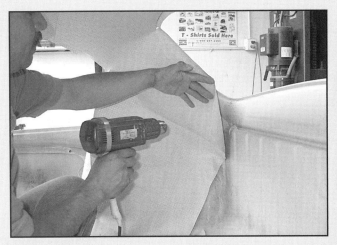

The most important step in making the rear of the headliner look good is maintaining the line that comes off of the bottom of the side windows. If your glue starts to set, heat it until it is tacky again.

From the back corner, mold the material over the rear wheelwells, following the curve all the way to the front. Cut away the excess material.

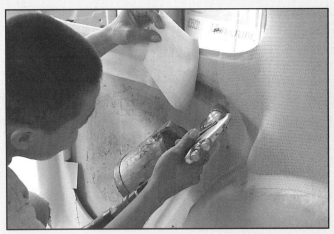

The final step for this kit is to attach the quarter window pieces and they merely glue into place without much trouble. Since they are the outside pieces, make sure the lines created are straight and even on both sides.

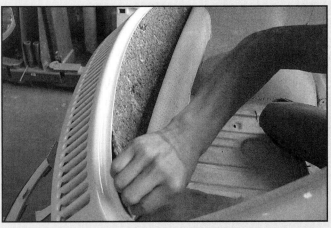

For later cars, especially Super Beetles, the package tray carpet kit is supposed to curve up and over this area and cover everything from the window down. However, we wanted the clean look of headliner material in this area instead of carpet, so Octavio fashioned this extra piece with sewn edges.

The bottom of which just tucks into the package tray channel and out of sight.

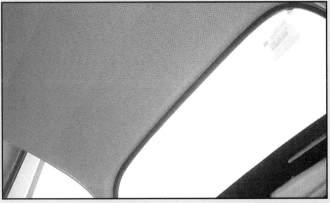

After about six hours of continuous work, the headliner is completely done and looking great.

INSTALLING RUBBER, DOORS, AND WINDOWS 20

Having a painted car back and finished in your garage is a great feeling and it is a turning point in the project, a milestone to signify the peak of the struggle to make this Super Beetle into a beautiful, yet functional, road-worthy Volkswagen.

We have finished the undercarriage, the running gear, the hidden mechanicals. We've scraped off the ugly undercoating, revamped the brakes and suspension, replaced all of the bushings, the transmission, the brake lines, the axles, the CV boots and a host of other things nobody will ever see.

Now we're ready for the fun stuff, the flashy stuff and the things that really makes building up a car interesting. In this chapter, we're going to take care of the door and window rubber, along with a couple of the outside seals.

For this step we took our Super to Classic VW Specialty's main man Rafael Gutierrez to show us the right way to rebuild the doors, replace the glass and seal it all with Wolfsburg West's rubber. As well, it is a good idea at this time to install the padded dash, as the top screws can be difficult to get in straight with the windshield in place. Also, the rubber from the windshield should butt right up against the

We started with a box of rubber parts from Wolfsburg West, a complete set of almost everything we'll need to protect our Super from the elements, from door seals, window scrapers to running board seals, door stopper seals and bumper guard strips.

pad...at least original dashes do.

There are very few specialty tools you'll need to tackle this step, and if you've got a 5/32 drill bit and a riveter (among the common tools you'd normally have) then you're out of excuses not to proceed on your own. However, if you're hesitant to take care of it yourself (or if there's any other step you don't think it wise for you to do), contact Classic VW Specialty as they are a do-it-all Volkswagen restoration shop.

Because you're working with glass, you'll need to create a safe work environment for your car. Get a flat workspace that's off the ground and cover it with a layer of foam or a couple of thick beach towels. That way, as you're manipulating the rubber over the glass, you won't run the risk of chipping the glass. Remember, chips equal leaks and leaks equal rust, mildew and mold, three things you don't want inside your car.

THE VOLKSWAGEN SUPER BEETLE HANDBOOK

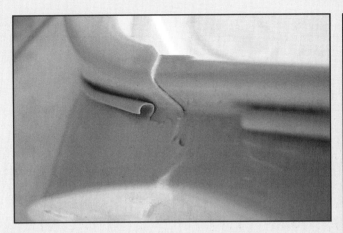

We started with the trunk seal, the outside rubber that runs the perimeter through this channel. The deck lid seal is installed the same way so we're going to skip it. When installed, crimp just the corners of these channel ends so the rubber stays in place.

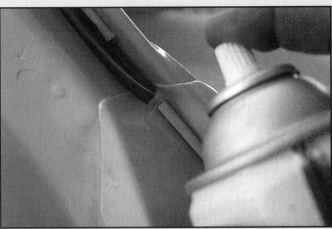

There are two ways to do it: The wrong way, which is to pry up the whole lip, slide in the seal and pound down the lip again, leaving it crimped and crumpled...and the right way, to feed the seal through the lip while keeping it well lubricated with silicone spray.

Feed in each side of the seal at the top of the trunk until both ends overlap at the bottom. At the top corners you'll have these attachment points that need to be carefully fed into the body holes. The best way to feed these through the holes is by using silicone and needlenose pliers. Tug on the thickest part of the rubber as possible, as these small pieces are fragile and tear easily. Squeeze as close to the body as possible and pull. You'll hear and feel a small pop when the rubber seats.

Moving to the windows, now is the chance to get them 100 percent clean, and since you'll hopefully never see the edges again, make sure they are very clean and free of any old rubber, dirt or debris. Any gaps will cause a leak.

Marking the center of the glass, the rubber slips over the glass with little effort. Make sure the center of the rubber pieces match with the center you've marked on the glass.

INSTALLING RUBBER, DOORS AND WINDOWS

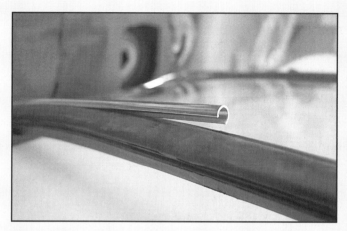

The chrome molding (actually, it's aluminum) is fed into a small channel in the middle of the rubber. It might be a little long at the end, so you'll have to cut off both ends to make sure it meets in the middle of the window.

While all of this is happening, the dashpad is installed. There are four screws on the bottom, two on the top and two studs that have to be bolted down with those special flanges we showed you in Part Two.

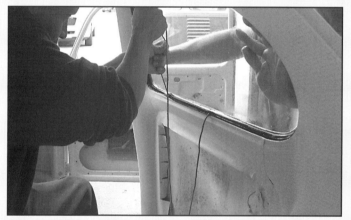

Once the rubber and trim are on the windows, feed a piece of wire or cord into the body groove of the window and give it a healthy dose of silicone. Some suggest soapy water as they feel petroleum products prematurely break down the rubber, but we like using silicone because it isn't as messy as soap and water. Have a friend push on the outside of the glass while you pull the string, in effect, pulling the lip of the rubber over the window lip and into the car.

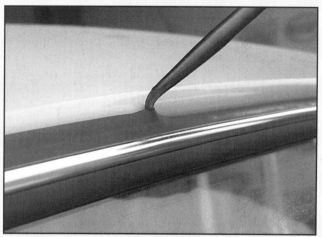

One specialty tool they have is this pointed screwdriver that pulls out the rest of the rubber over the car. Any curved piece of metal or a screwdriver will work, but be careful that you don't scratch any metal that will show.

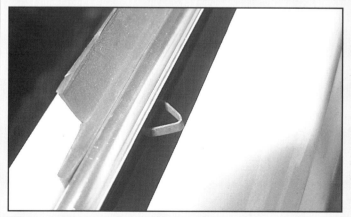

Now that the windows are in, our attention goes to the doors. The outer scraper goes in first (113 853 321). We've overexposed this picture to show you the clips that need to be lined up with the holes in the door frame. They simply snap into place.

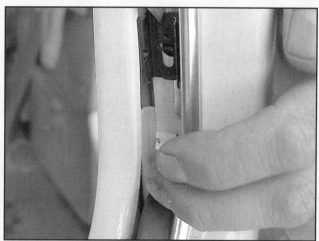

There are six of these clips for each door (111 837 361), three on top, two on the side and one inside the door. They help retain the felt channel the window rides in.

119

THE VOLKSWAGEN SUPER BEETLE HANDBOOK

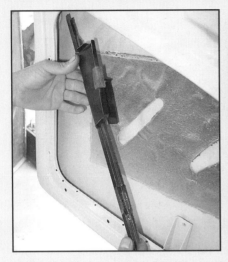 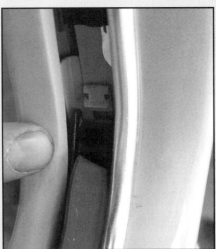

Up into the door goes the lower window channel. The channel screws into the door at the bottom but it snaps into place at the top of the door.

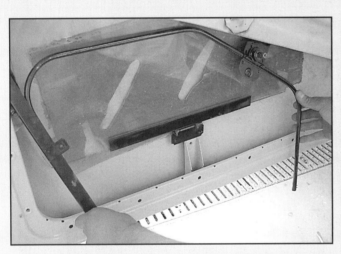 

Next, slide the glass into the door and set it in the back to make room for the regulator. We took a little time to clean off the dirt and gunk and give it a coat of black paint. Sure, nobody will see it, but we'll know it's there.

With a 5/32 drill bit, drill out the rivet that attaches the wing window to the frame. Take your time and don't drill away any of the frame around the rivet, otherwise your window will be forever loose.

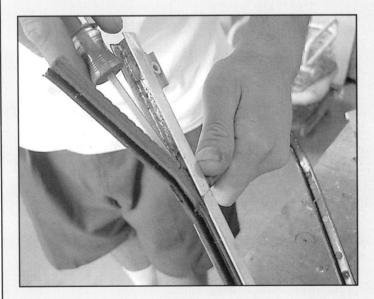

After unbolting the adjustment bolt at the bottom and the window comes out, scrape away the felt channel and the rubber molding. Make sure to clean the glass well and remove all of the glue from the channel.

INSTALLING RUBBER, DOORS AND WINDOWS

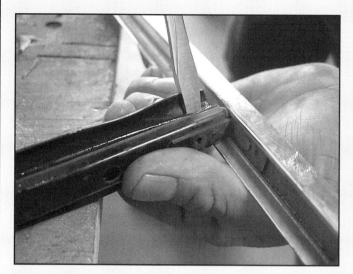
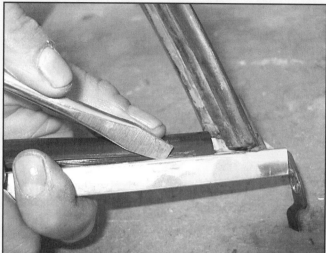

The new rubber is fed through the channel and pushed into place with a screwdriver.

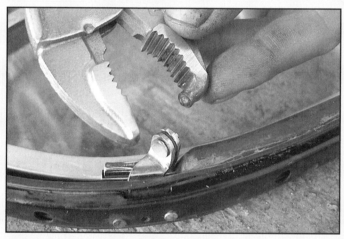

A new rivet is put into place and cinched down with a special tool that Rafael made himself from a pair of vice grips. Of course, a rivet gun will work all the same.

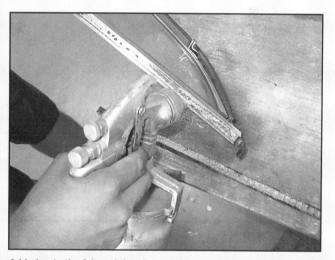

Add glue to the felt and the channel. Make sure to seat it properly in the channel so there are no tight spots or crimped areas.

The wing window assembly is fed into the door frame at a slight angle, then pushed into place. The top is held by a screw—make sure that the outside rubber is folded over the door. If not, slide something flat in there and pull out any stuck rubber.

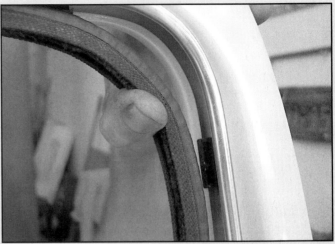

Next the main felt channel is squeezed into place by feeding it into the door. Start in the upper corner and feed the felt into the channel going both ways. Remember the six clips? They each have two spikes that hold this felt piece in place. Cut to fit on either side.

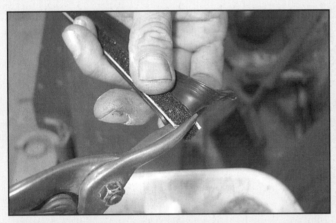

The inner scraper comes to us a little on the long side so the metal rail is cut along with a quarter inch of the rubber. Since every car is different, put yours in place before doing any cutting. Yours might just fit.

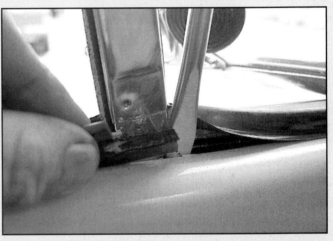

Where the scraper meets the wing window frame, you must lift up the wing window rubber and push the scraper end rubber into the door. This will lock the two pieces together.

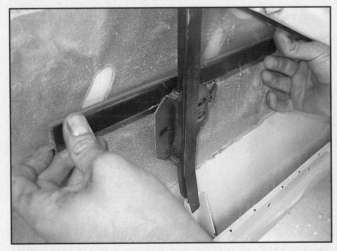

At this time, the window goes into place and gets attached to the regulator. Keep everything loose until you make the final adjustments.

Roll the window up and down several times, making adjustments on the regulator until you're comfortable with its action. With the window up, Volkswagen was good enough to add these holes into later years of Beetles to finally tighten the window to the regulator.

INSTALLING RUBBER, DOORS AND WINDOWS

Next is the door rubber. With German rubber, you need no glue, as it fits tight enough, but Brazilian rubber requires glue because it is harder and prone to slipping out. Start in the upper outside corner and work around the door.

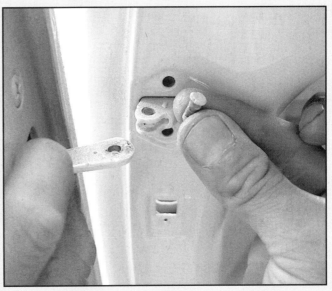

To complete the door rubber, the door stopper needs to be removed by knocking out this pin and freeing the stopper through the door. For correctness, the check rod shouldn't have any paint on it, so we wire wheeled ours down to bare metal (zinc coating would be appropriate). Don't forget the check rod seal on the door jamb.

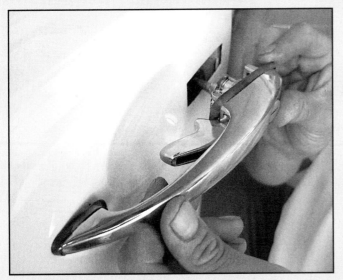

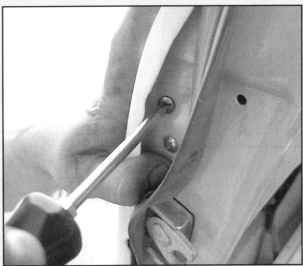

Since we had already installed the door mechanisms and the striker plate (they both simply screw into place), the new door handles from BFY Obsolete Parts merely slide into place (left). It is a good idea to scrape away the paint on the holes before fitting the handles, as it is a tight fit. Right: One screw for the door handle (the lower screw is for the mechanism) and you're set.

21 INSTALLING WIRING HARNESS AND ELECTRICAL COMPONENTS

The sight of a VW wiring harness can be very intimidating upon first glance. Uncoiled, it is 392 total feet of wires, 101 in all with 214 spade connections. That means there are 21,614 possible ways to install a wiring harness...but only one right way.

To make sure we got it right, we took our Super Project '71 to master electrician Rafael Gutierrez at West Coast Classic Restoration to show us the proper way to install the harness properly. He's done it so many times that a lot of his steps are second nature, making the whole process look deceptively simple. There are few specialty tools you'll need to make this work, as the basic ones found in any nicely equipped toolbox will do the trick. In addition, get a good copy of the schematics for your year. The '71 Super is one of the more complicated diagrams (as far as Volkswagens are concerned), but not as bad as the '73, which has the most wires, the most connections and the most components to deal with.

This might be perfectly obvious to you, but for safety's sake, don't connect the battery until the very end, and this is only after you have double- and triple-checked your wires, the connections and the proper operation of all equipment. Start by laying out all of your wires; separate each harness from the pile and sort the individual wires into groups based on size and location.

At West Coast Classics, our Super Project '71 was in good company. For most of its stay, the Super was flanked on the left by this beautiful '50 Split and on the right by a yet-to-be-finished Hebmüller.

INSTALLING WIRING HARNESS AND ELECTRICAL COMPONENTS

Don't worry about the mass of wires. Once separated and identified, each wire starts to make sense and the pile will recede to nothing.

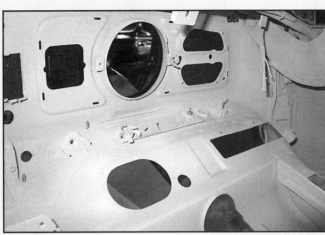

This is the blank canvas we had to work with, the freshly painted underside of the dash. There are a couple of extra holes you wouldn't expect on a Super Beetle, such as the cutout for the standard fuel gauge and the oval hole for the gas heater.

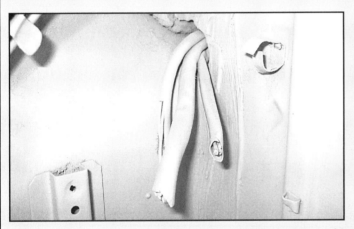

This is a very important step. When you're tearing out your old harness, make sure you don't remove the wires that go from the engine compartment to the left rear quarter panel. It is nearly impossible to reestablish this channel if the wires are pulled out.

Rafael starts any wiring project by taping the front of the main harness to the old wires in the engine compartment. Then, with the help of an assistant, pull out the old wires while pulling in the new harness.

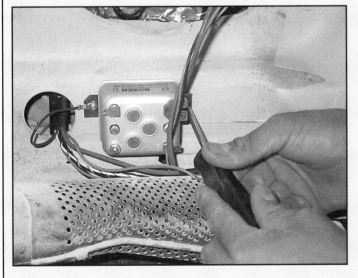

Once the main harness is pulled through, match up the last cutout with the regulator and attach the wires: #1 (connects to the fuse box via #75), #2 (to generator warning light on speedo), #12 (to the D+ terminal on the generator and #14 (to the DF terminal on the generator). On the left is wire #13, the ground.

THE VOLKSWAGEN SUPER BEETLE HANDBOOK

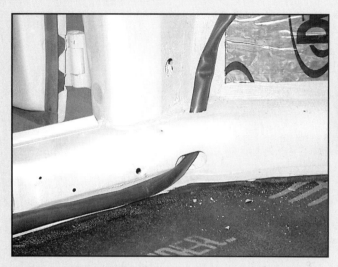 

Left: The front half of the harness is fed along the floor, under two clips and up through the front firewall. Don't worry, the carpet kit will cover these wires. Right: Don't forget this little rubber grommet on the "firewall."

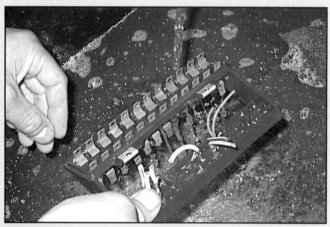 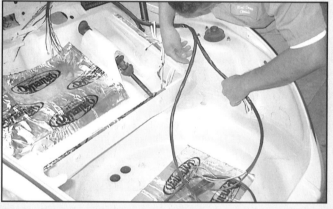

Since electrical components corrode as easily as metal parts, you'll want to give them a good scrubbing with solvent. Shown here is the fuse box with the attached relay console, a first for 1971 model VWs.

The front of the headlight harness consists of the headlight wires and the connections for the turn signals. The right side splits off and travels above the spare tire and under the fuel tank cross member. Farther back toward the dash on the harness is the cutout for the master cylinder harness.

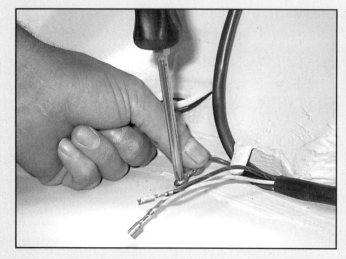 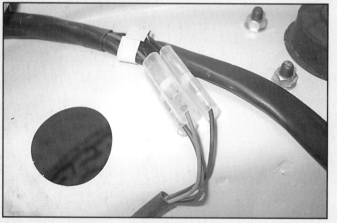

These are wires #17, #20 and #21 for the right-side turn signal. The ground screws into the body near the spare tire well, shown here.

This is the connection for the brake master cylinder. Though a harness is included in the kit for the brakes, we decided to reuse the old wires because of the original plastic connectors (much like the connections for the headlights).

INSTALLING WIRING HARNESS AND ELECTRICAL COMPONENTS

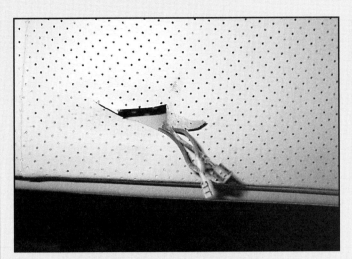

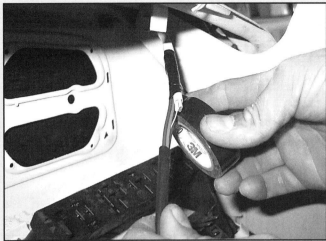

These are the original wires for the dome light. When the headliner was installed, a small slash was cut into the fabric to retain this location. Treat these wires similar to the main harness, as they must be attached to the new dome light harness and pulled through the A pillar.

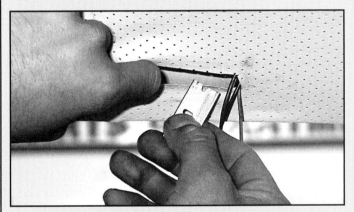

The headliner is cleaned up and a new dome light is installed. The red wire (#28) attaches on the left and ends up on the fuse box, while the brown wire with the white stripe (#30) connects from the right side to the door jam switches. The brown ground (#29) is attached to the speedometer ground.

With the old speedometer nearby, it is easy to transfer the bulbs from one to another, along with the plugs for the lights that won't be needed. Included on this is a new fuel sender vibrator which screws onto the fuel gauge.

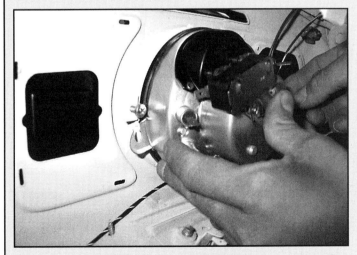

Probably the easiest thing we'll be accomplishing on these pages is the installation of the speedometer, which is attached with only two screws.

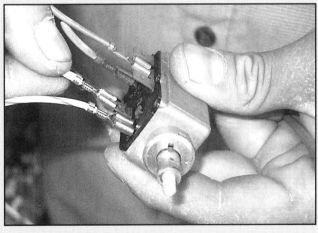

This is the headlight switch. Most of the wires can be attached while the switch is out of the car, but for the full setup, you'll need to install the switch.

THE VOLKSWAGEN SUPER BEETLE HANDBOOK

The headlight switch goes in the hole closest to the steering wheel, for easy access.

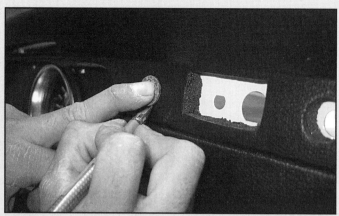

Because this is a general dashboard for most all Super Beetles, there is no hole here for the headlight switch in the pad, so one had to be cut. The reason is that 1972 and later Supers had the wiper switch on the column and the light switch is in its place.

The only special tool used to install the various switches is this wrench with a double pronged head. If you don't have one, use needlenose pliers instead.

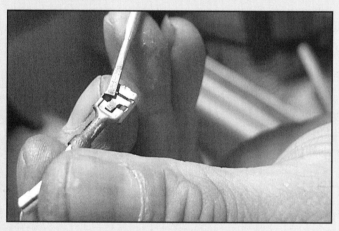

For most of the wires, if you are replacing the spade connectors, you'll need to clip off this retaining piece on the connector, except for those wires that attach into the relay switches. This clip holds them in place.

Sometimes it is easier to connect wires to the fuse box before it is screwed to the bulkhead. In this case, a group of wires needs to be routed from behind.

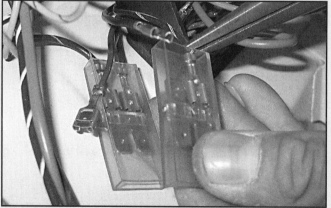

These are the quad connectors used to control the turn signals. They are attached this way so that power and function can act simultaneously, so your front and rear lights will flash at the same time. Wires in the connector on the left are for the right turn signals while the connector Rafael is working on is for the left turn signals.

INSTALLING WIRING HARNESS AND ELECTRICAL COMPONENTS

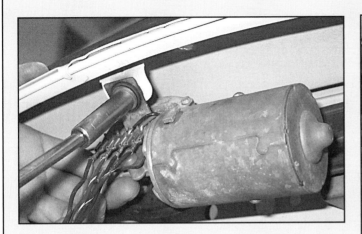

After a thorough cleaning, the wiper motor is returned to the body. It is a good idea at this time to replace the rubber grommets for the wiper posts. The wires for the motor are straightforward and easily accessible. With a 12-volt battery we tested to make sure both speeds of the motor functioned properly.

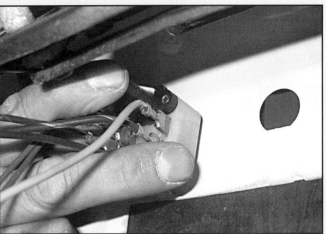

Behind it goes the wiper switch in the same manner as the headlights switch. The water tubes to the reservoir tank and squirter can be installed later.

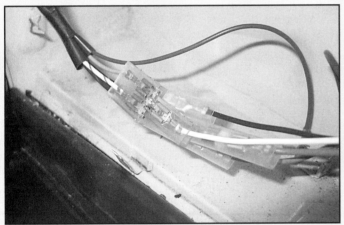

In the rear fenderwells are the wires for the brake lights, turn signals and reverse lights. The rear tail light harness is attached to the main harness via single and triple connectors. This whole system is hidden behind the engine's firewall material.

The silver ground strap will be connected to the negative terminal of the battery while the positive wire goes to the starter.

This silver box is the relay for the rear defrost. The wire to the actual window contacts had been retained (the front wire), but wire #42 on the left is the ground to the body while the black wire (#6) goes through the main harness to the defroster switch. The thick wire in the foreground is wire #52 that goes from the starter to the main harness (#11).

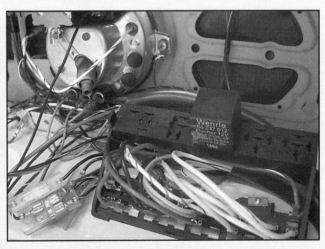

As you can see, progress is moving along behind the dash. If each wire is taken one by one, slowly they will fall into place.

THE VOLKSWAGEN SUPER BEETLE HANDBOOK

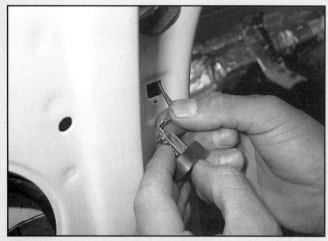

The door contact switches are installed in the door jambs. Normally on the left side is the buzzer unit that sounds when the door is opened while the key is still in the ignition. Since this is a most annoying sound it will not be included in our plans. Plus, original units rarely work after so many years.

Four screws hold the turn signal/horn contact plate.

This retaining clip holds the contact ring on the shaft.

Two small bolts hold the top to the ignition housing. Once the top is off, the ignition and steering lock slide out.

To make the new turn signal unit easy to install, the wires are splayed out then fed down the steering column. It is installed with the same four screws and spacers that were removed earlier.

Once the wires from the column are hooked up, the final product is a complete wiring harness. Since we still have a couple of details to take care of first (new headlights, turn signals and battery) we won't be testing the system until a later date.

SOUNDPROOFING OUR BEETLE

Our Super Beetle project is a shell of a car with nothing inside of the body but vibrating, clanking sheet metal. To combat this, we are installing sound deadening material from Dynamic Control, specifically DynaMat Xtreme and DynaMat Extremeliner. DynaMat Xtreme is a lightweight butyl material covered by a 4mm aluminum top layer that helps with heat (in addition to sound) resistance. DynaMat Extremeliner is a composite material with four parts: 1/8-inch of neoprene foam, 15mm of acoustic lead (a low-frequency barrier), 1/4-inch open/closed cell acoustic foam, and 3mm of urethane for a tough coating. According to Dynamic Control's Web site: "Extremeliner provides high acoustic absorption and excellent thermal insulation for a variety of automotive uses—from getting the most out of a mobile audio system to reducing low-frequency road noise to attenuating a noisy engine or drivetrain."

Installing DynaMat requires only a few tools: razor knife or scissors, roller tool (that you can get from Dynamic Control), rags and a solvent-based cleaner. When installing DynaMat on upside-down surfaces, you will need a spray adhesive. If heat is your main concern, first apply DynaMat Xtreme to the floor and firewall, then install Extremeliner, both underneath the floor's carpet for a double layer of protection.

In the end, we installed approximately 37 square feet of DynaMat material. We road-tested the car afterward and our sound level meter recorded an average of 58dB, with spikes only as high as 64dB, or about the same as an electric toothbrush. By the numbers, you could say that this is a 21 percent improvement, but since decibels are measured on a logarithmic scale, the DynaMat caused the intensity of the sound to be decreased 240 times.

The various kits we ordered for our Super Beetle project are shown here. The silver material in the foreground is DynaMat Xtreme, while the black in the background is DynaMat Extremeliner. Also included is the roller application tool, instructions and two DynaXorb kits for installation behind any speakers to redirect sound back into the car.

THE VOLKSWAGEN SUPER BEETLE HANDBOOK

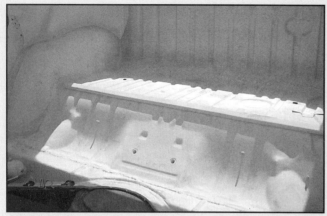

The first step to a successful application of DynaMat material is to completely clean all surfaces the sound deadener will attach to. We used BullFrog's Rust Blocker Cleaner-Degreaser and then rinsed it with a sponge and water. Here is the source of most of the noise that comes through a car, the rear package tray, a large expanse of sheet metal inches over the engine. When sound waves hit these panels, they vibrate and cause audible noise.

Sound and heat come up from the floorboards. On a hot day, the blacktop surface of the road can reach 130 degrees. And only a foot or so above that heat source, the pans can easily absorb and transfer the high temperatures into the car. As you're driving, wind bangs against the bottom of the car and causes the sheet metal to vibrate, transferring that vibration into sound.

Our door panels were a little tricky, as there was still some stock factory material stuck to them. Since it had been doing its job for so many years, there is no reason it can't continue to work. However, we'll cover the areas not covered by the stock material.

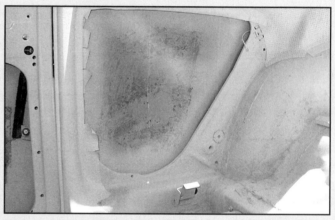

The rear quarter panel is a no-brainer. At 2.36 square-feet of sheet metal, it can vibrate like a kettle drum.

This is a shot of underneath the driver's side floor pan. As you can see we have already helped fight road noise by applying a couple of coats of rattle-can undercoating. It's not much, but it will add to the overall effect.

SOUNDPROOFING OUR BEETLE

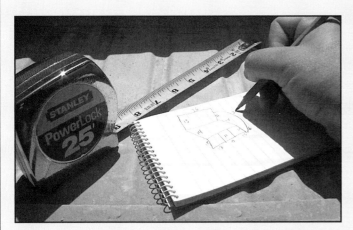

Back inside the car, we'll start our DynaMat installation with the package tray. Measure out the area so you'll know exactly how much DynaMat you'll need and what size pieces to cut. Dynamic Control suggests that you only need to cover approximately 50 percent of the surface, but since this area is a source of heat and a big source of sound, we want to cover as much as we can.

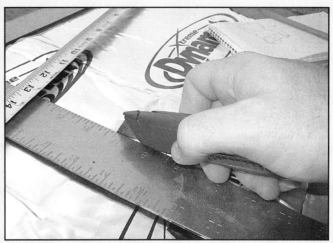

Transfer those measurements to the DynaMat Xtreme material. Use manageable pieces, as you can always overlap in layers if you need total coverage.

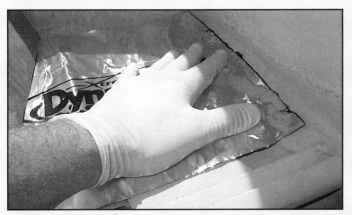

We thought latex gloves would keep our hands clean during the installation and would keep dirt and oil off of our cleaned panels, but the adhesive on the back of the material stuck to our gloves like gum. We soon ditched the gloves.

After you peel off the release liner on the back of the DynaMat, use the supplied roller to get rid of any air bubbles and to make sure the deadener completely adheres to the panel. For the rear package tray area, we used seven pieces and approximately 12.66 square feet of DynaMat Xtreme. For you weight conscious racers, at 0.45lbs per square-foot, that equates to only 5.6lbs to the overall weight of the car.

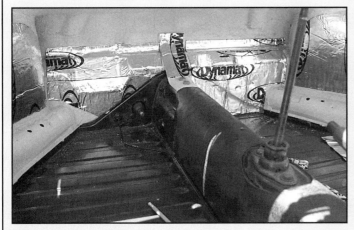

Since they provided us with roughly 60lbs. of Xtreme, we looked around the car to decide where else this stuff could benefit us. Basically, if something covered it, we started with a layer of DynaMat, shown here on the foot wells. We also added some to underneath the gas tank and spare tire well.

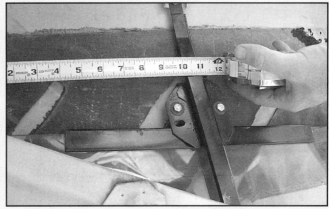

We measured the door for our application. Since there is original deadening material already there, we decided to merely layer around it, making sure not to affect the function of the window.

THE VOLKSWAGEN SUPER BEETLE HANDBOOK

We added three pieces to each door. This 8x12-inch piece, an 8x28 piece above this one and an 8x18 L-shaped piece to the right, notched out around the window regulator. This adds 3.625 square-feet, equaling approximately 1.631lbs. of weight to each door

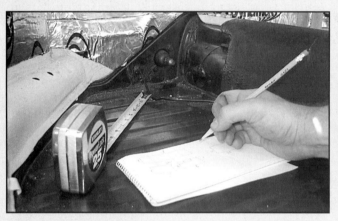

Next, we attacked the floor with the same method as the other panels, this time using the thicker and heavier Extremeliner. They suggest that you start with a layer of Xtreme underneath the Extremeliner and that's a good idea for extra protection.

Though thick, the Extremeliner cuts easily with a utility knife. For the driver's side, we cut a 28x17-inch piece, then notched out the needed areas. The pan narrows from the left 15 inches up from the bottom left corner and meets the top 8.5 inches from the top right corner. Additionally, we need to cut out a 9x4.5-inch patch from the top right corner to make room for the pedals.

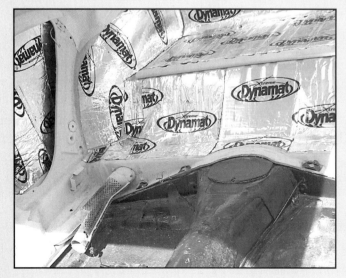

Once the backseat floor pans' 14x22-inch pieces were added (we used 3M adhesive), we had laid down approximately 9.83 square-feet of Extermeliner, and at 1.5lbs per square foot, that equals 14.75 additional pounds.

THE FUEL AND FRESH AIR VENT SYSTEM 23

Basically, there are no two ways about this: You can't install the fresh air fan and vent ducting system without installing the fuel tank and the evaporative emissions control system at the same time. There are a couple of steps that dovetail between the two installations and it all takes place in the close confines of the trunk. In addition to that, make sure all of your knobs are in place and that you've installed the dash-mounted ashtray. After the fact, it is nearly impossible. Not surprisingly, there isn't a wealth of information regarding the evaporative emission control system on the 1970 and later Beetles, except for the general explanation in the Bentley book and a short how-to about fixing a fuel leak at www.super-beetlesonly.com.

In short, the emission-control system consists of an expansion chamber that is connected to the tank that collects overflow gas from a full fuel tank. It is stored in this tube under the cowl until the fuel level eventually drops. In addition, the evaporated fuel from the tank (after the engine has been shut down) is forced up through the expansion chamber and into the ventilation lines. Then those fumes travel to the rear of the car where it is "washed" in an activated charcoal canister. The charcoal absorbs the vapors and, when the engine is restarted, the cooling fan blows the

Without the cargo liners (both top and bottom pieces) the completed trunk with the hoses and tubes should look something similar to this. Obviously the tank is in the foreground and the fresh air fan housing is at the top; however, there are a dozen or so parts that end up hidden in a variety of directions.

trapped vapors into the carburetor where the engine burns them off.

A note about safety: Old gas tanks, even if they seem empty, still contain vapors, and vapors plus an errant spark or a buildup of static electricity can cause a large explosion in a flash. To combat this, always disconnect your battery and ground your car. Keep a fire extinguisher handy, however.

Some details still need to be addressed. For example, we didn't connect the trunk release or the fuel door release levers. And we'll need to reinstall the correct fan shroud with the flange to accept the charcoal canister hose, and later we'll add to the system the water canister and lines for the windshield wipers, and consider replacing the jack.

If you smell fuel inside the cabin or inside the trunk, you've got a leak somewhere. Sometimes, the expansion tube and charcoal canister (which are no longer offered) don't work and must be accounted for. If that's your problem, you must reroute some lines to block them off. Avoid any leaks at all costs. In reality, a 1971 Super doesn't have to pass smog tests, and because of that, it doesn't have to be environmentally sound. Converting your tank to that of an earlier system is simple: connect a hose from the filler neck to the top left corner of the tank and connect together the two inlets on the upper right side of the tank, therefore bypassing the charcoal canister system and lines and the expansion chamber. For more information on this change, check out www.superbeetlesonly.com.

THE VOLKSWAGEN SUPER BEETLE HANDBOOK

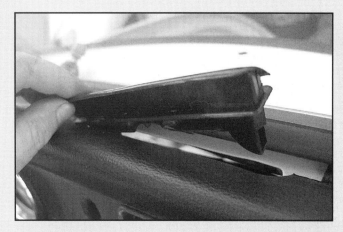

Starting on the inside of the car, snap the center vent (113-255-483) under the dash pad and onto the body. The dash pad might need a little alternation (use a clean razor) so the vent fits snugly on the pad. There are a few clips on the vent that snap under the dash.

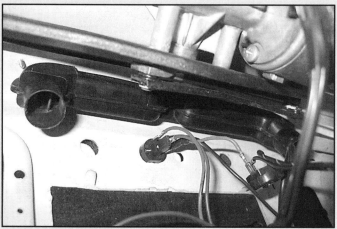

Underneath the cowling in that same location goes the center defroster vent housing, which they don't make anymore, so we hope you kept your old one. This piece snaps into the center vent and holds both pieces together.

These are examples of the hoses needed to connect the defroster vent to the defroster hoses in the quarter panel channels. On the right is the original plastic, ribbed hose, while the left hose is the one-inch paper aftermarket replacement piece. Though the plastic hose is better, as it needs to fit underneath the glovebox, the paper one is more forgiving in tight places and sharp corners.

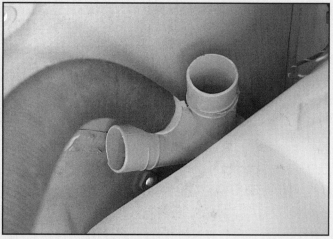

The vent hoses slide onto the center connection of the defroster hoses and wrap underneath the hood springs.

This is the original fresh air vent that directs air from the fan into the cabin of the car. This is another part that isn't currently reproduced so if your pieces aren't in good condition, you'll have to find replacements. All ours needed was a little cleaning and they were good to go.

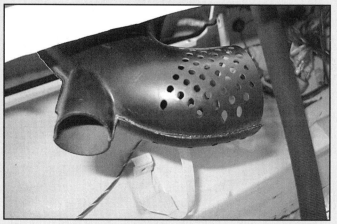

Once in the perforated tube heads to the fresh air fan while the other outlet is directed down to the defroster hoses in the quarter panel channels. On the dashboard are the corner defroster vents (113-819-635A) which snap into place similar to the center vent.

THE FUEL AND FRESH AIR SYSTEMS

Because our original vents that fit into the face of the padded dash crumbled to dust when we removed them, we needed to find a donor dash to take them from. Luckily this dash came from a cut-up '71 Standard Beetle. A razor blade helped cut the tabs out of the padded dash.

At this time it is a good idea to install the steel ventilation line up under the cowl where it is held down by three tabs on the body.

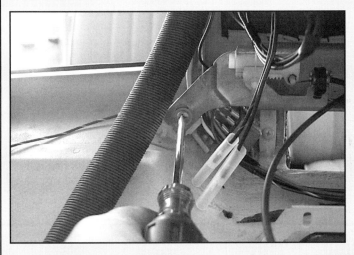

We fed the fresh air control knobs through the dashpad and secured the mechanism with these two screws. The control wires should criss-cross over each other twice so they don't bind during operation.

These corrugated plastic pipes, officially known as fresh air box hoses (113-819-717B), connect the corner vent housings with the fresh air fan. Feeding these pipes deep into the fan housing allow for a tight connection within the system and will enable you to drop in the fresh air fan box.

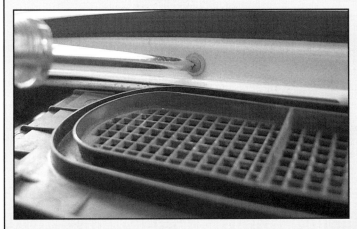
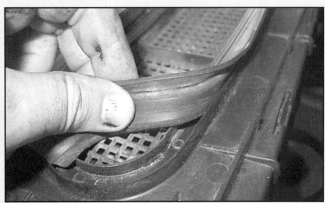

The fresh air fan box slips under the cowl and is secured via three screws and washers. To create an airtight suction that won't draw in air from the trunk, this hood seal (113-819-519C) fits neatly into the channel.

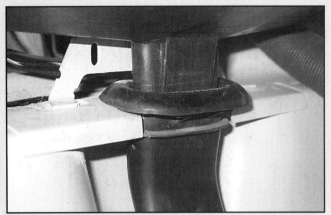

There are two types of drain hoses for fresh air fans. This one, 113-819-533, fit Super Beetles, while the other two-piece units fit standard Beetles. Also, we are missing the strap that holds down the bottom of the fresh air box.

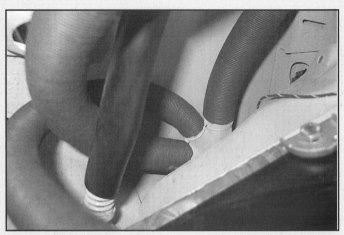

With all of the vent hoses attached, each side looks something similar to this. Although we haven't plumbed each hose where it needs to run, we do have them hooked up correctly. In this picture, the left hose runs to the side vent, the center hose runs to the center vent and the right hose goes up into the A-pillar.

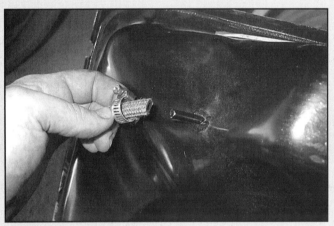

Once the vent system is mostly completed, we can begin work on the fuel tank. First start by attaching approximately 12 inches of fuel hose onto the tank's outlet tube with an appropriately sized hose clamp.

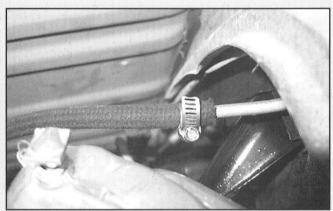

Underneath the tank, the 12 inches of fuel line is fed onto the hard line in the front frame horn. Of course, we used a properly sized hose clamp.

Back up top, we fed the filler neck into the tank and gasketed it with a filler neck seal (111-809-599A) and then this filler neck hose (311-201-219A). Two large clamps hold it tightly in place.

Once the filler neck is secured onto the tank it needs this rubber gasket around it to keep the filler neck from vibrating on the body.

THE FUEL AND FRESH AIR SYSTEMS

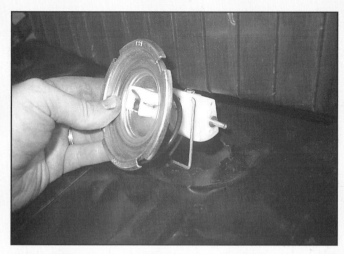

Finally, the fuel sending unit is fed into the tank. The unit can only attach into the tank in one direction, and it tightly cinches down over a rubber gasket.

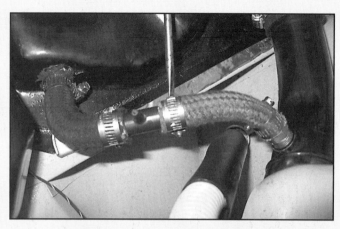

Let the connection of the hoses begin with this collection of fuel lines that run from the filler neck back into the top of the tank. Within this line is a T-connector that allows fuel vapors to enter the expansion chamber. Appropriately sized hose clamps are used throughout.

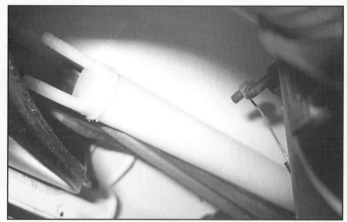

The plastic expansion chamber sits under the cowl in this area. Though there's no particular place to bolt it down, it should fit snugly on top of the windshield wiper motor so it doesn't interfere with its movement.

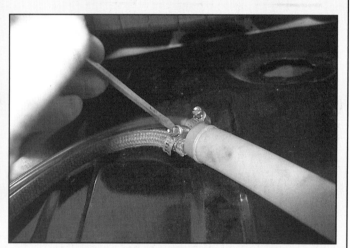

On the passenger side of the expansion chamber (closest to the filler neck) is connected a length of 3/16-inch (inside diameter) clear plastic tubing and a 16-inch piece of fuel line.

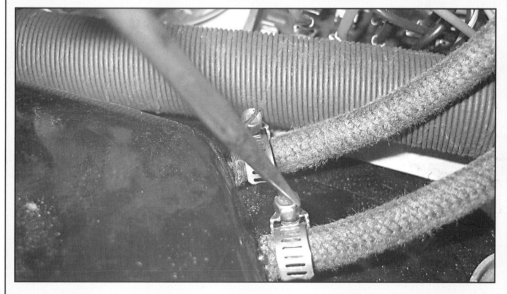

On the other side of the expansion chamber (driver's side) are connected two 12-inch pieces of fuel lines. Those two lines attach to the fuel tank and serve as a conduit to re-circulate overflow fuel back into the tank.

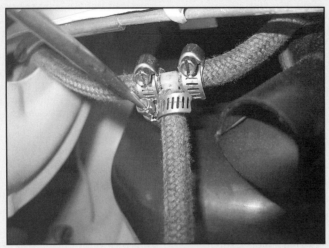
Cut into the fuel line leaving the passenger side of the expansion chamber is a T-fitting and another 12-inch piece of fuel line that connects to the T-fitting in the fuel filler neck line. Confused yet?

The plastic tube from the expansion chamber joins the steel ventilation tube located under the cowling.

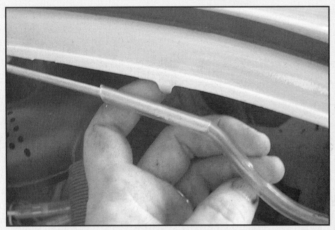
On the other side of the ventilation tube is connected another length of plastic tubing that feeds down the same channel next to the fuel tank that the hard lines for the brake reservoir feeds through.

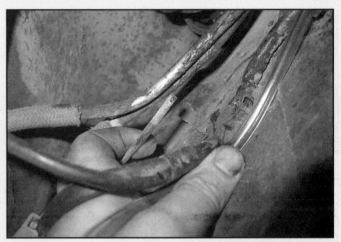
The plastic line that connects the ventilation tube to the "firewall" cross piece does so underneath the car near the brake lines.

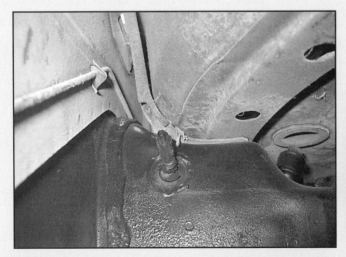
Down underneath the car on the "firewall" is a cross piece of steel ventilation line that inexplicably returns to the passenger side of the car.

The firewall steel ventilation line is connected via fuel hose to another steel ventilation line that runs the length of the pan. Missing are the hose clamps that we will add later.

THE FUEL AND FRESH AIR SYSTEMS

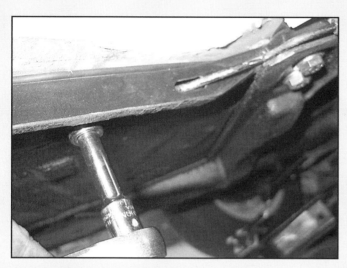

The long line attaches to the car via mount flanges that connect underneath three to four body bolts.

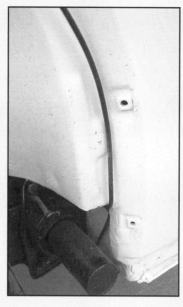

Underneath the rear passenger fender is yet another piece of steel ventilation line that connects to the long line in a similar manner to the front "firewall" piece. This line is held in place by the fender bolts, and then connects to the bottom of the charcoal canister with a foot-long fuel hose.

The thick hoses from the canister are fed into the engine compartment. The top hose connects to the fan shroud while the bottom hose wraps around the back of the fan shroud (some feed it through the deck lid spring) and connects to the air cleaner on top of the carburetor.

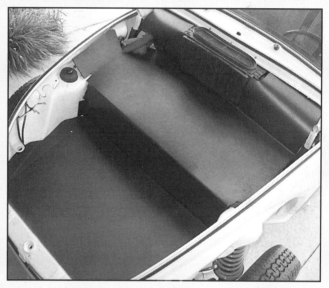

Back to the front of the car, we did some minor adjustments to the two-piece trunk liner, namely we were forced to cut clearance space for the fresh air fan box. Once installed we considered the job done.

24 FINISHING THE INTERIOR

The carpet is one of the finishing touches of any custom interior, and shouldn't be taken lightly. We contacted Wolfsburg West in Corona, Calif., to have them install their 11-piece carpet kit.

When you are considering a carpet kit, you must take into consideration that there are two kinds of carpets used on the 1970-and-later cars (Super and Sedan). The combination of carpets and rubber mats were stock for these cars, but they came with two different kinds of material, both considered "original." The first is the highest quality 36-inch-tight square weave carpet, which is considered as the best. In addition to this was a lesser grade material called Perlon; though not plush, it was original. The price is the same for both of these carpets, but for our 1971 Super Beetle, we decided upon the square weave carpet (Part No. 113-863-014) because we wanted the best. Also, we knew that our Super was sold as a well-equipped model, therefore we concluded that it would have come from the factory with the higher-quality carpet. Each and every piece is finished off by a sewn-in twill border.

Underneath it all should be plenty of padding, and that can come from a variety of sources. Stock is Wolfsburg West's three-piece carpet pad kit (113-863-891) for the rear package tray which consists of two layers of jute covering sandwiching an asphalt center.

Now, installing the carpet kit behind the rear seat isn't possible without recovering the backseat rest first, which is a consideration if you're redoing all of the upholstery, as we did. To install the carpet kit, there aren't many special tools you'll need. We show you a glue-gun air-tool, which most people don't have. Instead, use any glue designed for upholstery, such as 3M's Super 77 Spray Adhesive. Other than that, equip yourself with a couple of clean razor blades, a putty knife and a collection of basic tools.

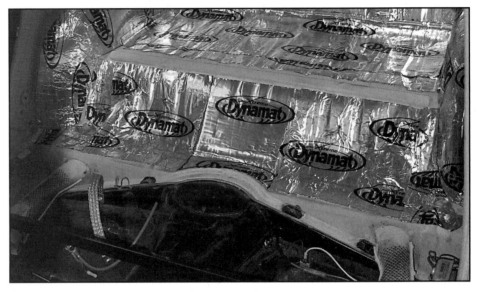

This was the state of the interior when we wheeled the Super into the shop. We had installed the two layers of DynaMat Xtreme and DynaMat Extremeliner as soundproofing material in Chapter 22.

UPHOLSTERY

There are two types of seat covers, one from a box and one custom made. The one from the box is great, assuming it fits, but nothing mass produced is going to fit perfectly, so some adjustments are necessary. We've used seat cover kits before in our project cars, but they just don't have the same unique appeal of a custom interior. However, it is very difficult to make such an interior yourself unless you have access to all of the industrial equipment necessary, so it is not a project to be undertaken lightly. We can't possibly teach you how to make a custom interior in this chapter; it would require a whole book. HPBooks does publish one, *Auto Interiors*, if you'd care to tackle it yourself. See page 170 for details on ordering.

So, we turned our car over to Octavio Kustom Upholstery in Orange, CA and had him work his magic on the interior.

FINISHING THE INTERIOR

A thin layer of rust was removed from the passenger footrest and the whole thing was given a nice coat of black paint, even though no one will ever see it again.

Before you consider laying any carpet (or putting on the panels), you've got to clean up your old parts that you've decided to reuse. On that list for us is the foot well vents, the door pull plates and most of the hardware. Start by removing any dirt that has built up.

To make the parts look brand new, use a plastic/vinyl prep coating and a flexible black paint. Remember, applying a couple light coats of paint is better than one thick one.

While those parts dry, spray a thick coat of adhesive on all surfaces inside the car that will be covered with carpet. Over spray can easily be removed by either gasoline or any solvent. Allowing the glue to get tacky (after about 15–20 minutes) it will stick better.

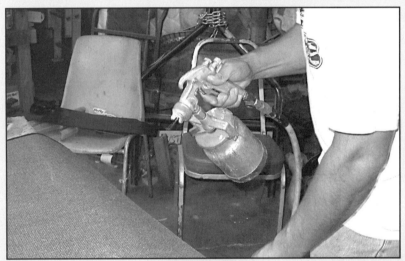

Start by spraying another thick coat on the three carpet pieces that will fit on the rear package tray and the fenderwells.

The top curve must line up so that it folds down the wheelwell in the front while curving around to the vertical side where the package tray meets the fenderwell.

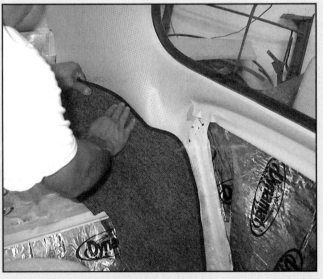

The top curve must line up so that it folds down the wheel well in the front while curving around to the vertical side where the package tray meets the fenderwell.

FINISHING THE INTERIOR

With a piece of cardboard to block the spray, apply a coat of glue to the edges of the wheel well pieces. Watch for overspray.

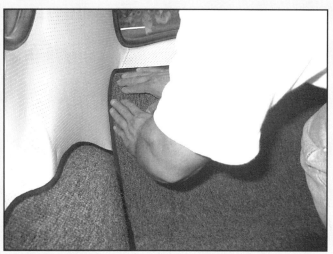

Since the main package tray piece has been tacking up, it should be ready to be applied to the tray. It should overhang across the top of the package tray's lip by an inch.

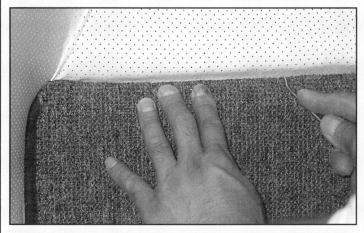

The edge of the carpet is folded down into the channel behind the package tray. To fit it into the corners, we needed to cut off approximately an inch triangle of material at the corners. The tool in Octavio's hand is a thread puller used to keep the threads lined up and straight.

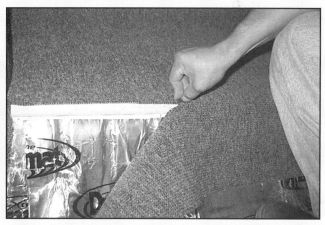

The rear main carpet kit needs to be cut to fit. We cemented the top portion first and then cut it long enough so the bottom of the carpet would fit into the carpet channel that is behind the back seat. Once cut, it can be tucked into the channel with a putty knife.

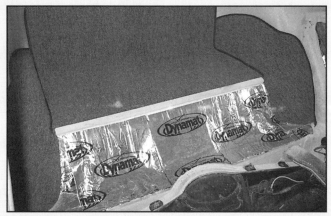

Here is the final product behind the backseat. Once that is done, it may have taken you a little longer than Octavio, so you'll have to add another coat of adhesive.

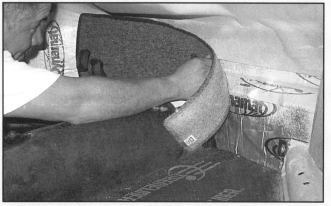

The same sequence happens for all the other carpet pieces. First we place the foot well piece, and we did this piece first to make sure that it is centered on the "firewall" and not affected by the placement of the other pieces, which actually go under this front piece.

THE VOLKSWAGEN SUPER BEETLE HANDBOOK

Next are the threshold pieces, one under each door. These both have a rubber lip that hooks over a metal channel running along the bottom of the door jamb.

This metal channel must be pried up slightly to make room for the rubber flashing. Don't worry about the paint chipping or cracking, as it will be covered by the rubber.

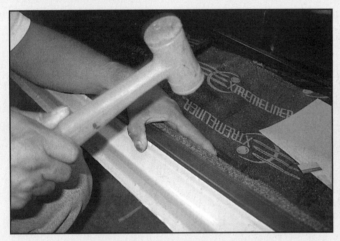

Once in, straight and smooth, tuck the carpet piece down next to the seat runners, pulling it tight. Flatten the metal channel with a rubber mallet.

Since this carpet fits a wide variety of years, holes need to be cut to accommodate the foot well heater vents. Make these cuts with a new utility blade.

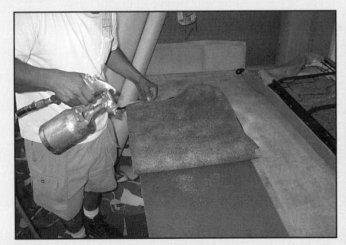

The last pieces to go in are the kick panels. These pieces actually fit under the threshold and firewall pieces. It is at this point where the backseat rest needs to be completed if you're going to proceed.

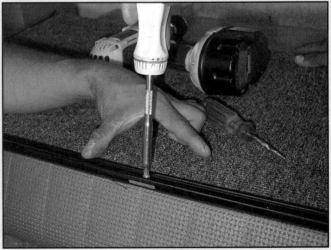

Once the carpet piece is glued down to the backer board, the seat rail is screwed into place, starting with the center screw.

FINISHING THE INTERIOR

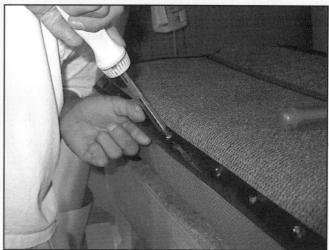

On top of, but facing it, is this extra piece of carpet that covers the hinged area of the seat when it is folded down. The metal strip keeps everything in place. It is secured with five screws.

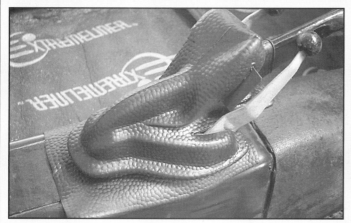

Before fitting the rubber mat (or floor carpet piece if that's what you've ordered), make sure the boots for the parking brake and the gear shifter are in place.

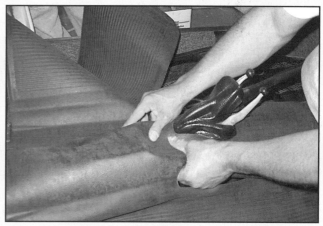

Fitting the rubber floor mat is pretty straight forward. We decided to glue it to the tunnel to make sure that it doesn't bubble up and/or affect the operation of the gas pedal.

The final carpet kit installed.

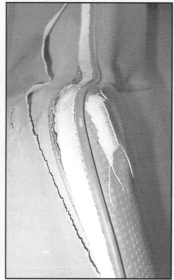

In the 15 years these seats have been sitting outside in the ever-changing elements, time, heat and moisture have caused the old covers to split open, and interesting enough, the "old" covers were merely covered over the original from-the-factory seat covers the car was delivered with.

147

THE VOLKSWAGEN SUPER BEETLE HANDBOOK

Pry the backrest from the base, and pull off the old cover (covers in our case) from the frames.

Teardown is the easy step. Save the thin rods from the base and mind the spikes at the bottom of the backrest.

The knob for the back seat release simply pulls off as well as the plastic molding that surrounds the lever.

The seat cover can then be pulled off the frame.

This is a nice shot of all of the original elements that make up a Super Beetle front seat. The sisal, the horsehair padding, the foam headrest, the springs, the cardboard backing and the plastic moisture barrier will all be removed.

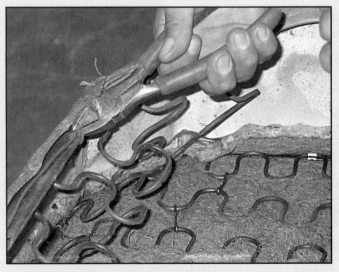

The backseat is torn apart via the same methods. There is a long thin rod that must be retained for the rebuild.

FINISHING THE INTERIOR

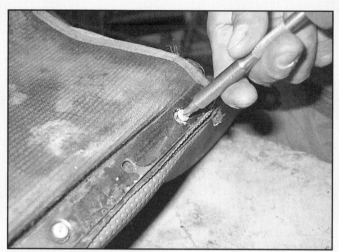

The back seat backrest contains a couple of things you'll need to hold on to, such as the top rail and the metal strip that holds the carpet to the back seat.

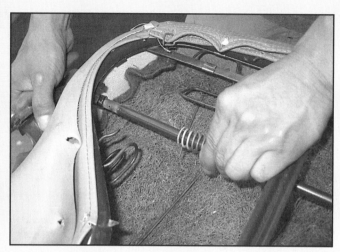

The backseat release mechanism must be removed as well, and this is done by unclipping the spring. It is held in place with a small pin that can be driven out.

Once the seat frames were cleaned and fixed, the first step toward revitalization is the addition of a layer of sisal, a 1/4-inch-thick piece of composite cloth material that keeps errant springs from piercing the newly completed seats.

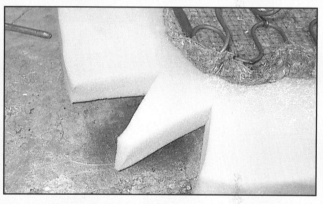

Starting with the back seat, a layer of glue is sprayed on the sisal and a layer of foam padding. An M is cut into each corner of the padding so that it doesn't bunch up when folded over the seat frames.

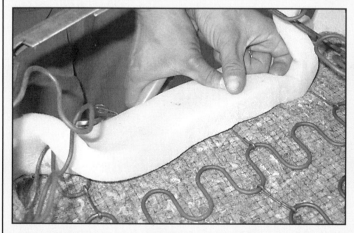

Like the sisal, the padding is hog ringed to the frame. Hog ring pliers can be found at most any VW parts house as well as any automotive upholstery shop. Keeping the sisal tight to the frame will make for a solid seat that doesn't easily sag in the middle like poorly rebuilt seats.

For the backrest of the front seats, the sisal is cut to only cover the springs and not curve around the back. This is so there isn't any extra "over-stuffed" padding on the sides of the frames.

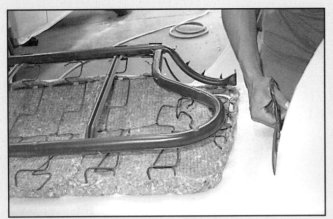
By contrast, the padding for the backseat backrest is cut a few inches larger than the frame so that there is padding around the back side as well.

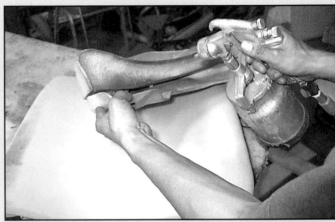
Glue is applied to the foam headrest and the top three inches of the backrest foam padding. We decided to reuse the original headrest as there was nothing wrong with it...and we knew it would fit the frame easily.

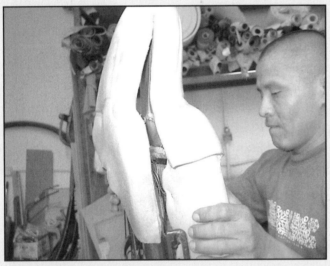
The headrest should be tight on the frame and over the padding of the backrest.

Once the headrest pad is secured to the frame and the top of the back rest padding, a bit of glue is applied between the seams and it is held down for a few seconds.

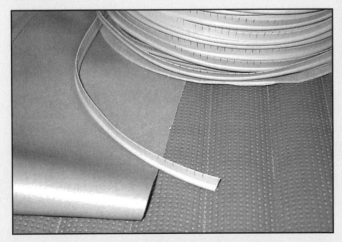
The needed vinyl material comes in a basket weave and a smooth pattern, much like it did originally 35 years ago. These are the only raw materials—approximately five yards each—you'll need to complete the job.

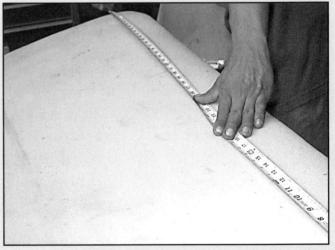
After the foam padding has been applied to the seat frames over the sisal, the edges are marked, measured and re-measured. If there is even a slightly miscalculated measurement the whole seat cover will look awkward.

FINISHING THE INTERIOR

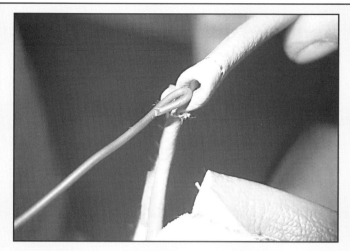

Once the seats are together, you can easily replace the wire that runs through the front and the back of the panels. If you've lost the wire, anything with the thickness of a coat hanger wire will do. Make sure to curl the edges around like this so as not to puncture the seats.

While the front seats are being sewn, the rear bench seat is being planned out. The pleats are laid out by marking three-inch spaces in chalk. The center point on the materials is right between the middle pleat. This will keep the pleats on the seat to match up with the seat backrest.

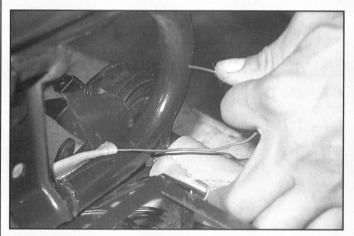

Not only are the wire-enforced seat panels locked down by the metal spikes, but the wire that runs the perimeter of the bottom of the seat is tied off on the frame.

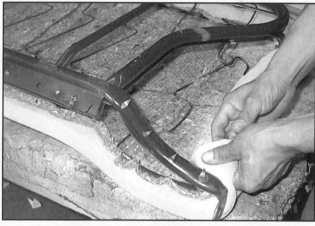

The one-inch seat foam is glued to the sisal as well as hog ringed to the frame.

Now that the pleats have been laid out, the sewing begins, allowing for slight shrinking after compressing the padded backing. Notice that the padding is cut slightly larger than the vinyl material. This can be cut away later.

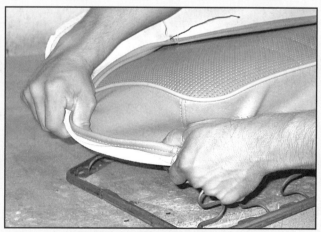

If it isn't a tight fit, then it isn't going to be a good fit. That is why measuring is such an important step in any part of this process. Turn the cover inside-out first and stretch it over the frame. Vinyl has quite a bit of stretch, so don't be afraid of pulling on it.

The wire that was strung through the bottom pleat helps keep the edges tight and smooth. Tie it off around the frame. That same wire also adds a secure spot to hog ring the cover to the frame. If the vinyl rips at all during its life, it won't break through the wire.

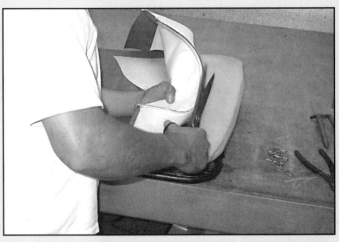

We noticed on the backseat backrest that the sewn edges on the cover (under the welting) made the cushion look bulky, so some of the extra was cut off.

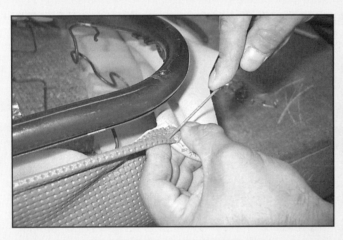

This is a close-up shot of the wire being feed into the pre-sewn channel along the bottom of the backseat backrest. It serves a similar function as the one for the bottom cushion.

The backseat backrest has several spikes that help hold the vinyl in place. The spikes are sharp so in order to keep blood off of your new seats, avoid spiking your palm.

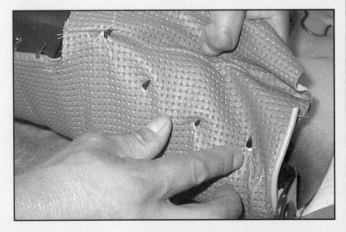

These four spikes secure the underside of the cushion that curves up over the wheelwell. Then secure the cover in the same fashion as the others.

For best results, try not to rely on the precut backer boards you'd get from the kits. They are usually either too big or too small (mostly too small), so a new one is measured by tracing around the original one.

FINISHING THE INTERIOR

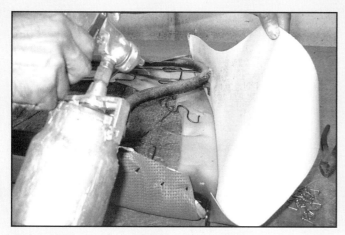

We jumped back to the construction of the backseat backrest. Here, glue is applied to the side flap and allowed a few minutes to set up.

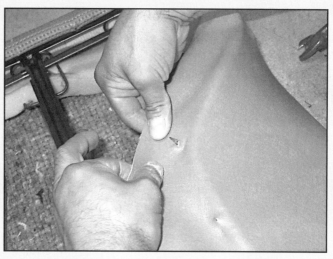

It is folded over and pulled down over the set of spikes on the vertical frame member.

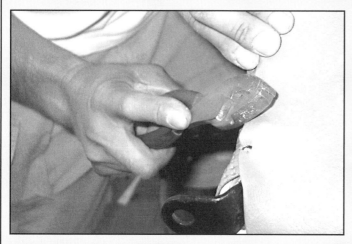

What is left over on the bottom is folded up into the seat and pulled down over the same spikes seen earlier.

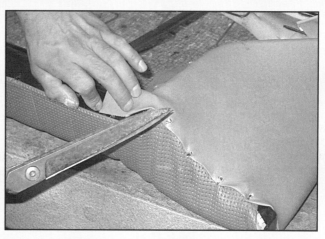

The leftover pieces are cut away so they don't bunch up and create a ripple that the backer board must form over.

The backseat seatrest release bar spans the entire backseat allowing the release mechanism to pull the levers on both sides of the seat at the same time.

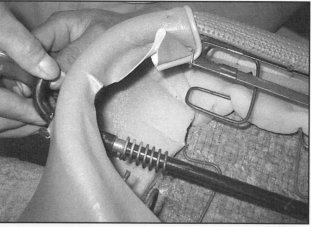

The backseat seatrest release bar slides through a hole cut into the side panel and through to the other side. The spring is held into place by a small pin that must be replaced.

153

THE VOLKSWAGEN SUPER BEETLE HANDBOOK

Now jumping up to the front seats, apply a generous amount of adhesive to the corners and the headrest padding.

For some extra cushioning, we've applied another one-inch-thick piece of padding to the corners.

Again, to install the newly sewn covers to the front seats, turn the covers inside out and start at the top. Make sure the covers are properly lined up, as this part of the job takes a lot of tugging and pulling.

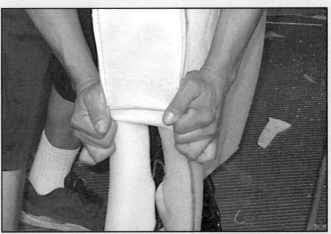

It is best to push down on one side a little and then switch to the other side, only sliding a little bit at time until the cover is completely tight at the top.

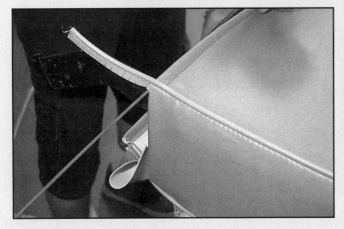

On the bottom of both sides of the seat cover we slide in a metal wire similar to the wires on the previous seats. These, like the others, add stability and strength to the edges.

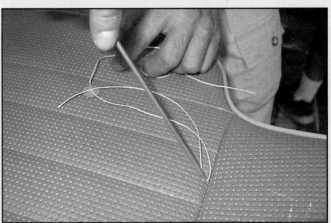

This long "needle" is used to insert the string that holds together the front and rear tufting buttons. There are two on each side, eight in all.

154

FINISHING THE INTERIOR

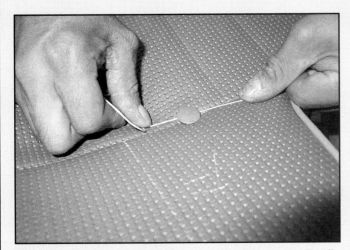

The original used wire, but a tough string will work perfectly. Tie off the front with the string in the eyelet, push in the back of the seat slightly to tie off the back tufting button.

This is the end result after three days of work at the sewing machine. We finally have a beautiful interior to match a beautiful car.

In the backseat, you'll notice that the pleats on the bottom cushion and the back rest match perfectly. The piping is straight and the covers are tight without looking lumpy or wrinkled.

THE VOLKSWAGEN SUPER BEETLE HANDBOOK

The door panels were specially constructed and recovered by Octavio's expert crew.

It is difficult to find them, so if you've still got the ashtrays (even if you don't smoke), include them now. After a little buffing and polishing, our old tarnished ones (one on each side), were as good as new.

Once assembled, the door panels are a perfect match with the interior, and we are happy to add that no other car has this same pattern

FINAL INSTALLATION OF EXTERIOR BODY PARTS AND TRIM

This is really where the car starts to near completion. Provided we had a battery and the coil was hooked up to the distributor we could drive away, knowing we have a perfectly restored 1971 Super Beetle.

But we're not there yet. We still have a few final parts to install, but the end is near.

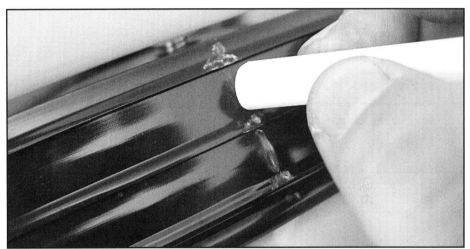

We started by attaching the fenders and the fender beading. Each fender is hung with all of the bolts loose. The beading comes in one long piece with no pre-cut notches, so they had to be made. We cut the beading into four roughly equal lengths

Laying the beading down the length of the fenders we marked with chalk the location of each bolt, 10 each on the front fenders and eight each on the rear. Then we cut a roughly one-inch notch at each chalk mark.

THE VOLKSWAGEN SUPER BEETLE HANDBOOK

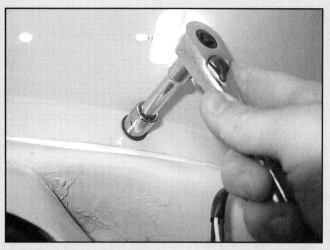

Once the beading was in place, a 13mm socket wrench tightened each fender in its proper place.

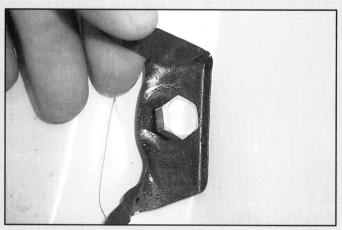

On the rear right fender, six of these ear-tabs are incorporated into the fender bolts to hold the rear fuel vapor line that routes its way to the carbon canister mounted nearby.

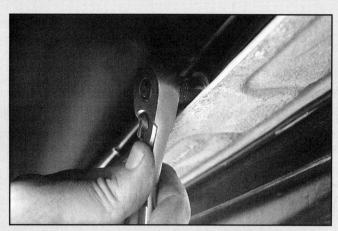

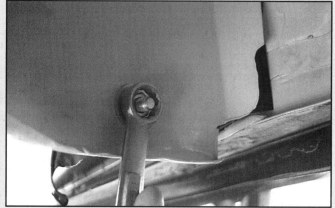

Next we'll tackle the running boards. Four 10mm bolts hold the board to the body (left photo), while 13mm wrenches on each side of the fender bolt the running board to the fenders (right). Actually, the picture on the right is incorrect, as we put the head of the bolts on the fender sides to make for a cleaner look. The rubber grommets go between the fenders and the running board.

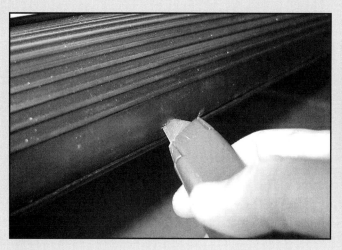

The chrome trim isn't already attached to the running boards, so we have to do it. Start by locating the mounting holes and feeding the trim clips into the trim pieces, five each. Slip the clips into the holes and, with pliers on the underside, twist the tabs to secure.

FINAL INSTALLATION OF EXTERIOR BODY PARTS AND TRIM

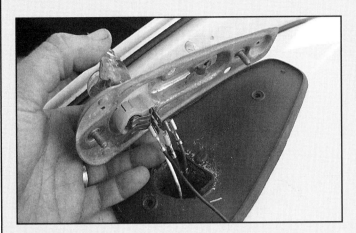

We fed the wires into the rubber boot by wrapping them together with electrical tape and dousing the inside of the boot with silicone lubricant. The wires slipped easily through. Once secured via two 8mm nuts, the new lens and chrome housing is screwed down.

The same goes for the boot that holds the wires for the headlight. There are three Phillips screws that hold the headlight into the bucket, and these are used to adjust the beam. One screw holds the chrome headlight ring to the fender.

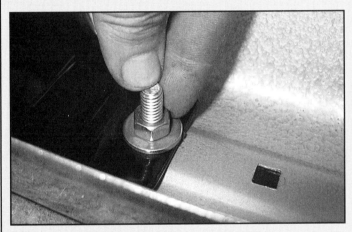

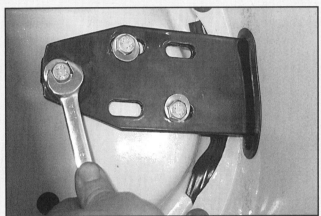

We had the bumper brackets powder coated, and there is a difference between front and rear, as there is a difference between the front and rear bumper. Three bolts hold each bracket to the bumper. Three 13mm bolts keep the brackets tight with the body.

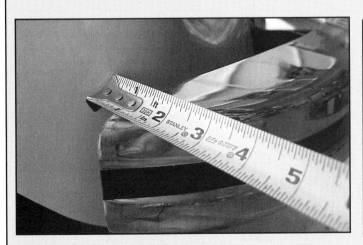

The brackets have oval holes for adjusting the bumper, and we used a tape measure to keep the distance from the bumper to the fender equal on each side, roughly an inch and a half.

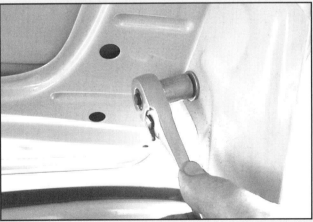

The real trick is to attach the hood by yourself. It involves a little contortionism and a strong back. Rest the hood on the cowl and with your left hand, hold the front of the hood until the holes line up. Use your right hand to start the bolts. Better yet, get a friend to help you.

THE VOLKSWAGEN SUPER BEETLE HANDBOOK

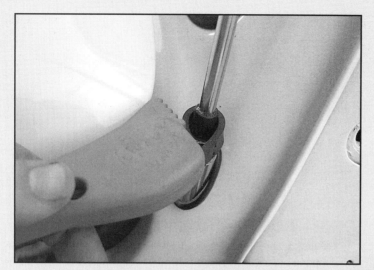

Even though we didn't install a radio (and probably won't) we didn't want to fill the antennae hole with anything but an antennae. The chrome top ring unscrews, and you install the antennae up through the cowl. Slide the top ring over the antennae and tighten it with a strap wrench so you don't mar the chrome.

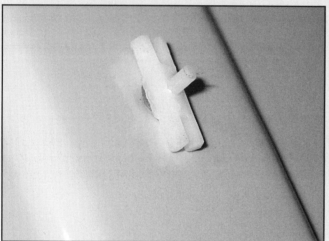
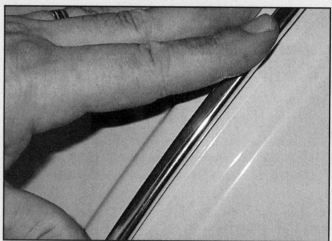

This is the plastic chrome body trim clip, and you'll need about 25 of them. First make sure to clean out the trim holes with a utility knife. The small tab on top of each clip must be pushed in to keep the clip tight. A slight tap with a hammer does the trick.

The all-important VW logo badge drops into place and is secured by three small rubber tabs. Since they're clear, be careful you don't drop them!

FINAL INSTALLATION OF EXTERIOR BODY PARTS AND TRIM

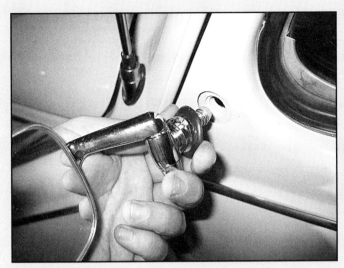

One of the easier things to install is the driver's side mirror, as it merely screws into place. Don't forget the rubber grommet.

The hood spring took a little bit of figuring out, but when you realize that the spring needs to be relaxed when the deck lid is up, you'll soon get it right. Clip the rear part on first and then pull the clips to the decklid.

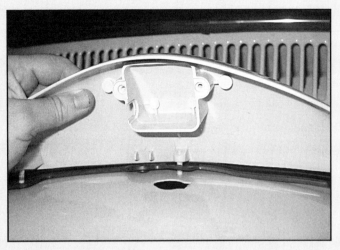

The decklid is attached with four 11mm bolts, and the decklid can be adjusted in and/or out by moving the bolts up or down on the brackets. If you haven't removed the brackets, you should have no trouble returning your decklid to its former position. After it is on, install the license plate hood and rubber trim. Three 8mm nuts fasten the hood to the decklid.

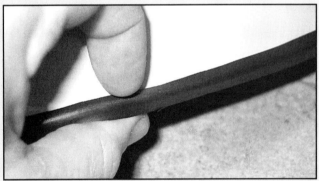

One of the more difficult things to do was to fit the rubber boot around the taillight housings. When you get one side on, the other side slips off. Patience is a virtue.

Up underneath the rear fenders are the 7mm bolts required to attach the taillight housings.

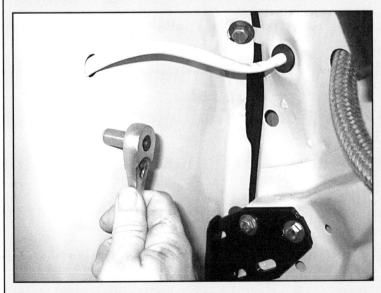

THE VOLKSWAGEN SUPER BEETLE HANDBOOK

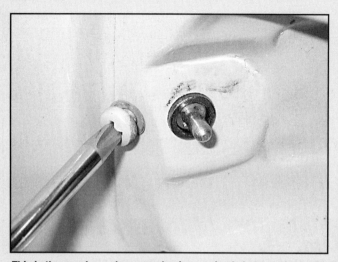
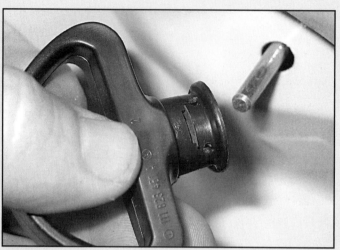

This is the gas door release mechanism, a simple button release attached to a short cable that goes into the cabin. One Phillips screw holds it down. Inside, the handle slides over the end and is held in place by a small clip that you must first push into the handle.

Two 10mm bolts hold the door to the body.

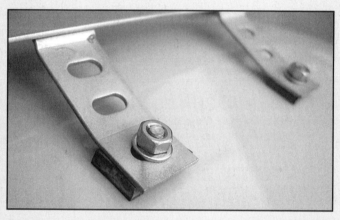

Back to the rear of the car, we attached the license plate bracket like this. Again, we placed the bolts "backwards" so that when you look into the engine compartment, you'll see bolt heads instead of nuts and washers. The rubber squares are to keep the bracket away from the new paint job.

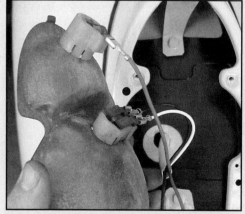

The taillight bulb housings are wired according to your wiring diagram, exactly like you see here. Two small screws fit the bulb holder to the housing.

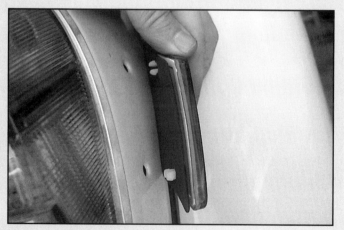

Once the taillight lenses are screwed in along with the aluminum trim (we used our old ones because we couldn't locate any new ones on such short notice), merely snap in the side reflectors.

FINAL INSTALLATION OF EXTERIOR BODY PARTS AND TRIM

We hope you kept your aluminum trim that fits around the rear quarter window vents, because they don't make them with the trim. Another road block is that the original trim won't fit on brand new vents. These are our old ones, but they are missing some of the fins. Odds are good we're going to go with the new ones without the aluminum trim.

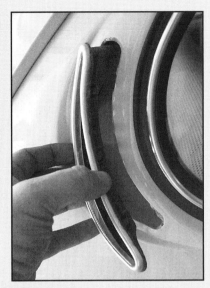

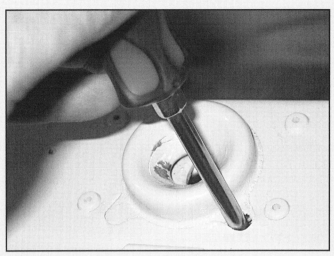

The cable for the hood release slides into the mechanism and is held into place with this screw.

On the other end is the release handle that fits into the glovebox. Since we didn't have a riveter, this step will have to wait, but the idea is there. Run the cable (and the sheath) around the right side of the trunk down to the release mechanism.

Up on top are the two wiper posts. An added detail is the black plastic bolt cover that sits under the wiper arm.

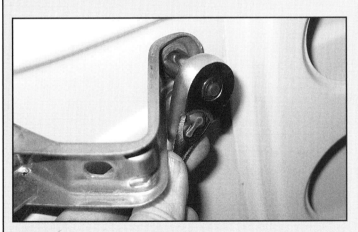

Next are the hood handle, the release mechanism and the rubber grommets. Note we didn't install the spring and catch because we haven't sorted out the cable system yet. There's nothing worse than locking your trunk with no way of opening it again.

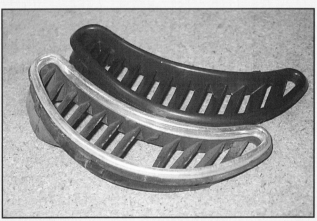

Details, details. We still have a few areas to finish up. We covered the rear vent fins earlier, but we installed the original ones. On the top is the aftermarket one, but it doesn't come complete with the aluminum trim, and the trim from original ones won't fit on the new ones.

163

As you can see, the flanges of the aluminum trim slide into channels cut into the plastic. The new vents don't have such channels. Even if ones could be fashioned, the trim won't fit over the molded plastic.

After all of that, the new vents merely snap into place on either side. There is a right and left.

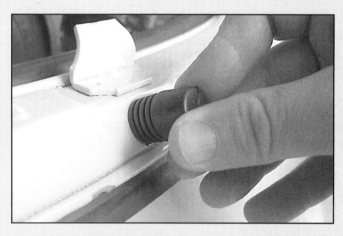

This is the rubber stopper that screws into the bottom of the decklid lip. It keeps the deck lid slightly away from the body to prevent rubbing and eventual paint loss.

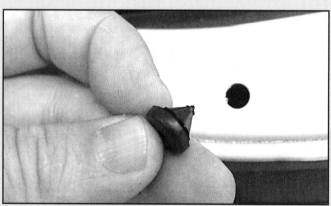

On either side of the rubber stopper are these small grommets. They serve a similar purpose as the main stopper. By adding these and the above stopper, the deck lid might need a little encouragement to closer securely, but we'd rather have a tight-fitting deck lid than a rattling one.

Like we said earlier, we didn't have access to a riveter, so we had to wait, but when we were at Octavio's Upholstery, he easily snapped three rivets on the trunk release for us.

In addition, we had to remove the plastic aftermarket glovebox to feed the two hinges into their receiver holes on the outsides of the glove box.

FINAL INSTALLATION OF EXTERIOR BODY PARTS AND TRIM

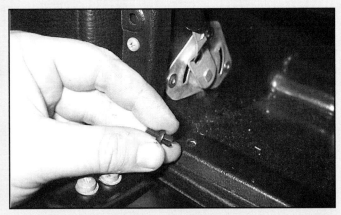

One of the problems we ran into while fitting the plastic glovebox trim piece to the padded dashboard were these little plastic clips. They are similar in design to those used on the outside chrome trim, but they weren't strong enough to hold the trim in place. We ended up using the original screws instead. We blame the padded dash as it had a little extra padding around the glovebox area.

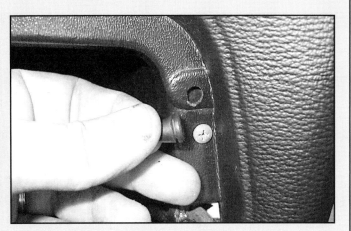

Once the screws were in on the trim piece, they were strong enough to push the padding deeper into place. We fitted the rubber stoppers to keep the glove box from rattling.

Finally, the locking glovebox door knob easily fits into place and is secured with the nut (provided you kept it!).

The two-piece dash grilles fit through the four slots on either sides of the speedometer. Originally, underneath the grills should be a piece of mesh and a layer of leather (or vinyl), but we decided to skip those until we found suitable material.

On the back side of the dash the four tabs are then pulled and twisted to keep the grille tight against the padded dash.

Underneath each front fender, we fed the headlight wires into the rubber boots, but we ran into a problem with the wires for the horn, which come out of the same hole in the body as the headlight wires. In order for the boot to have a tight fit on the body, we cut a slit in the side of the rubber and fed out the horn wires.

165

THE VOLKSWAGEN SUPER BEETLE HANDBOOK

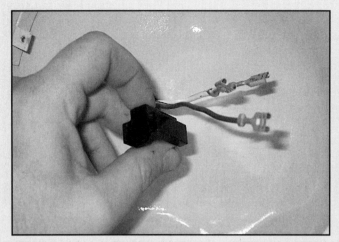

These are plastic inserts that help feed the wires into the headlight housing. The white wire goes on your left, the brown on top and the yellow plugs into the right. This holds true on both headlights.

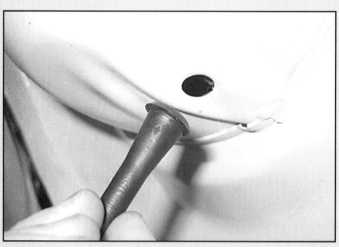

These are small rubber tubes that drain water from inside the headlight buckets, which catch a lot of water as you drive through rain and puddles or when you wash your car.

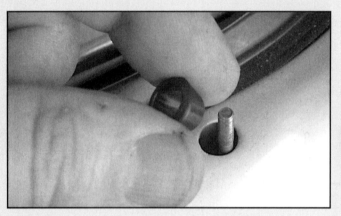

When we were rebuilding the doors a few chapters ago, we forgot to include these small collars that keep the door lock knobs from rubbing on the freshly painted doors.

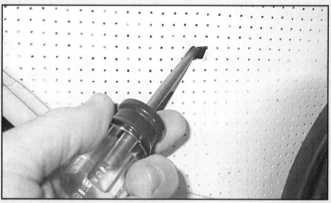

We couldn't remember exactly where the holes were for the sun visors, so we were forced to do a little exploratory surgery to find the main hole and the two screw holes on one side and the two holes for the hangers in the middle. One little trick is to shine a flashlight up onto the headliner and look for the dark spots.

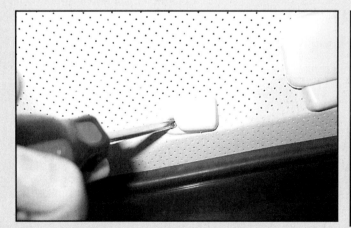

Once discovered, two small holes are cut into the headliner with an awl and a single screw holds the sun visor hanger.

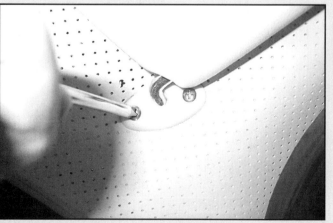

The same thing is done on the mounting point for the sun visor. Be careful with the screwdriver or you'll end up with the mar in the headliner like we did. Note the small hole.

FINAL INSTALLATION OF EXTERIOR BODY PARTS AND TRIM

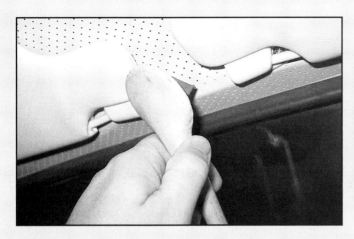

Of course this is where the rear-view mirror goes, but you'll have to cut about a one-inch X in the headliner. First make sure you've got the proper place, because if you don't you'll cause the headliner to stretch awkwardly. It actually snaps into place and is designed to break away easily in an accident. The mirror of the 1971 Super Beetle has a white stalk with a black backing, but the only aftermarket units we could find were all black. That wouldn't do, so a stock unit was found. Though dirty, it cleaned up nicely.

For the rubber bumper guard strips, there are two small rectangular holes in the bumper that the metal flange fits into. Then the strip is tightened by two screws on either end.

Finally done, the Super Beetle is rolled out of the garage to pose for the camera. It sits slightly high on all the corners because it hasn't yet been driven to settle the shocks and springs. Once it has covered a few miles, things will fall into place.

TECHNICAL SPECS

ENGINE AND TRANSMISSION
1302: 77mm bore x 69mm stroke, 1285cc
Compression ratio 7.5:1
Solex 30 PICT-3 to 6.71 then 31 PICT-1
44bhp (DIN) at 4100 rpm
Final drive ratio 4.375:1

1302 S/LS: 85.5mm bore x 69mm stroke, 1584cc
Compression ratio 7.5:1
Solex 30 PICT-3 to June 1971 then 34 PICT-3
50bhp (DIN) at 4000rpm
60bhp (DIN) at 4400rpm, US December 1970
Final drive ratio 4.125:1

SUSPENSION AND BRAKES
Front: transverse link axle, coil spring and MacPherson strut with dual circuit discs on S/LS (drums US), drums on 1302

Elements of the suspension, both front and rear, were also utilized in the 1976 Porsche 924. (Some '71, and to a lesser extent '72, models were manufactured with torsion bar/ball joint front suspension for the North American and South African markets.)

Rear: semi-trailing arm with double-jointed drive shafts and drums

WHEELS AND TIRES
4x15 rims (part no. 141-601-025) with 5.60 x 15 up to 112-2490-158 (2.72) then 5x15 (part no. 111-601-025F) with 6.00 x 15 up to 112-2961-362 (7.72)

DIMENSIONS
Wheelbase: 7' 11.75" / 2420mm
Front track: 4' 6.25" / 1379mm
Rear track: 4' 5.25" / 1350mm
Length: 13' 4.5" / 4080mm
Width: 5' 2.5" / 1585mm
Height: 4' 11.75" / 1500mm
Curb weight: 1918lbs / 870kg
Convertible: 2028lbs / 920kg
Front luggage capacity: 260 liters (from 140)
Rear ventilation slots 1971 models, three on each side for 1972
Fuel tank capacity: 42 liters / 11 U.S. gallons
Fuel rating (all models): 91 Octane

PERFORMANCE
1302: max speed: 125 km/h (78 mph) / 0–50 mph—14s
Automatic: max speed- 120 km/h (75 mph) / 0–50 mph—16.5s
1302 S/LS: max speed: 130 km/h (81 mph) / 0–50 mph—12.5s
Automatic: max speed- 125 km/h (78 mph) / 0-50 mph —14.5s
0-62mph: approx. 23s

SPARK PLUGS
Bosch W 145 T1, Beru 145/14,
Champion L88 (L88A from 1.3.71)

SEAT BELTS
3 (6) points to front, 3 (6) points in rear

BODY/CHASSIS REVISIONS
Brake discs enlarged by 5.5mm—June 1971
Engine lid contour modified and vents increased from 10 to 26, rear window size enlarged at top by 4cm, new hood and cover for Convertible, catalytic converter added for export models to California—August 1971
VW Diagnosis maintenance test socket to engine compartment—June 1971
Enlarged US front turn signal incorporating parking light and reflector

U.S. taillight assembly with back-up light and side reflector
All US models—buzzer/ignition key warning, red side reflectors to rear taillight assembly, side marker lights and reflectors incorporated in enlarged turn signal/parking lights. Back-up lights made standard—August 1971
Export models to U.S. in 1970 incorporated a Hella made reflector (VW part no. 113-945-103A) bolted on to the bumper bracket.

INTERIOR REVISIONS
Safety steering wheel, steering column wiper/washer lever—August 1971 Sedans only—rear interior ventilation slots increased from 2 to 3 with flaps, rear luggage space cover added—August 1971

FACTORY OPTIONS
1200 34bhp engine and front discs (1302 only)
Semi-automatic transmission
5.5x15 Sports wheels (part no. 111 601 025J) with 34mm back spacing (5.72)
Metallic paint
Sliding steel roof
Heated rear window
Eberspächer BN-4 fuel-electric (auxiliary) heater
Tonneau cover (Convertible)
Padded dashboard cover
High-back (bucket) seats
Non-dazzle interior rear view mirror
Wiper delay switch and intermittent wash-wipe device—August 1971
"L" package (standard on Convertible:
   back-up lights, rubber inserts in bumpers,
   padded dashboard cover, door pocket,
   two-speed fan powered ventilation,
   dual circuit brake warning
   light (as part of fan switch assembly),
   lockable glove compartment lid,
   make up mirror in passenger's sun visor,
   non-dazzle interior rear view mirror,
   two ashtrays in rear

U.S. DEALERSHIP ACCESSORIES
Air Conditioners
Luggage Rack
Whitewall Tires
Radial Tires
Rear Speaker
Solid Walnut Shift Knob
Sport Steering Wheel
Bumper Overriders
Sports Wheel Covers
Simulated Walnut Dashboard Kit
Stripe Kit
Trailer Hitches
Cocoa Fiber Floor Mats—August 1970
"Formula Vee" package since 1970 models
Bosch driving lamps for Baja Champion SE—1972

TECHNICAL SPECS

SUPER BEETLE SEDAN MODEL INFORMATION

Year	Production	Model	Engine	Engine Code	Chassis Code
1971	Aug. 70–Jul. 71	All	1600	AE0000001–AE0558000	1112000001–1113200000
1972	Aug. 71–Jul. 72	US/Can	1600	AE0558001–AE0917263	1122000001–1123200000
1972	Aug. 71–Jul. 72	Calif.	1600	AH0000001–AH005900	1122000001–1123200000
1973	Aug. 72–Jul. 73	Manual	1600	AE0917264–AE1000000	1332000001–1332212117
1973	Aug. 72–Jul. 73	Man/US/Can	1600	AK0000001–AK0239364	1332000001–1332212117
1973	Aug. 72–Jul. 73	Man/Calif	1600	AH0033404–AH0101888	1332000001–1332212117
1973	Aug. 72–Jul. 73	Auto	1600	AH0005901–AH0114418	1332000001–1333200000
1974	Aug. 73–Jul. 74	Man/US/Can	1600	AK0239365–AK0239493	1342000001–1342999000
1974	Aug. 73–Jul. 74	Man/Calif	1600	AH0101889–AH500000	1342000001–1342999000
1974	Aug. 73–Jul. 74	Auto	1600	AH0114419–AH500000	1342000001–1342999000
1975	Aug. 74–Jul. 75	Manual	1600	AJ0000001–AJ0059664	1352000001–1352600000
1975	Aug. 74–Jul. 75	Auto	1600	AJ0000001–AJ0012405	1352000001–1352600000

SUPER BEETLE CONVERTIBLE MODEL INFORMATION

Year	Production	Model	Engine	Engine Code	Chassis Code
1971	Aug. 70–Jul. 71	All	1600	AE0000001–AE0558000	1512000001–1513200000
1972	Aug. 71–Jul. 72	US/Can	1600	AE0558001–AE0917263	1522000001–1523200000
1972	Aug. 71–Jul. 72	Calif.	1600	AH0000001–AH005900	1522000001–1523200000
1973	Aug. 72–Jul. 73	Manual	1600	AE0917264–AE1000000	1532000001–1532212117
1973	Aug. 72–Jul. 73	Man/US/Can	1600	AK0000001–AK0239364	1532212118–1533200000
1973	Aug. 72–Jul. 73	Man/Calif	1600	AH0033404–AH0101888	1532212118–1533200000
1973	Aug. 72–Jul. 73	Auto	1600	AH0005901–AH0114418	1532000001–1533200000
1974	Aug. 73–Jul. 74	Man/US/Can	1600	AK0239365–AK0239493	1542000001–1542999000
1974	Aug. 73–Jul. 74	Man/Calif	1600	AH0101889–AH500000	1542000001–1542999000
1974	Aug. 73–Jul. 74	Auto	1600	AH0114419–AH500000	1542000001–1542999000
1975	Aug. 74–Jul. 75	Manual	1600	AJ0000001–AJ0059664	1552000001–1552600000
1975	Aug. 74–Jul. 75	Auto	1600	AJ0000001–AJ0012405	1552000001–1552600000
1976	Aug. 75–Jul. 76	Manual	1600	AJ0059665–A0095935	1562000001–1562200000
1976	Aug. 75–Jul. 76	Auto	1600	AJ0012406–AJ0023504	1562000001–1562200000
1977	Aug. 76–Jul. 77	All	1600	AJ0095936–AJ0119687	1572000001–1572200000
1978	Aug. 77–Jul. 78	All	1600	AJ0119688–AJ0132850	1582000001–1582050000
1979	Aug. 78–Dec 79	All	1600	AJ0132851–AJ0149558	1592000001–1592043634

Other Books by HPBooks

GENERAL MOTORS
Big-Block Chevy Engine Buildups: 1-55788-484-6/HP1484
Big-Block Chevy Performance: 1-55788-216-9/HP1216
Camaro Performance: 1-55788-057-3/HP1057
Camaro Owner's Handbook ('67–'81): 1-55788-301-7/HP1301
Camaro Restoration Handbook ('67–'81): 0-89586-375-8/HP1375
Chevelle/El Camino Handbook: 1-55788-428-5/HP1428
Chevy S-10/GMC S-15 Handbook: 1-55788-353-X/HP1353
Chevy Trucks: 1-55788-340-8/HP1340
How to Build Tri-Five Chevy Trucks: 1-55788-259-2/HP1259
How to Hot Rod Big-Block Chevys: 0-912656-04-2/HP104
How to Hot Rod Small-Block Chevys: 0-912656-06-9/HP106
How to Rebuild Small-Block Chevy LT-1/LT-4: 1-55788-393-9/HP1393
John Lingenfelter: Modify Small-Block Chevy: 1-55788-238-X/HP1238
LS1/LS6 Small-Block Chevy Performance: 1-55788-407-2/HP1407
Powerglide Transmission Handbook: 1-55788-355-6/HP1355
Rebuild Big-Block Chevy Engines: 0-89586-175-5/HP1175
Rebuild Gen V/Gen VI Big-Block Chevy: 1-55788-357-2/HP1357
Rebuild Small-Block Chevy Engines: 1-55788-029-8/HP1029
Small-Block Chevy Engine Buildups: 1-55788-400-5/HP1400
Small-Block Chevy Performance: 1-55788-253-3/HP1253
Turbo Hydramatic 350 Handbook: 0-89586-051-1/HP1051

FORD
Ford Windsor Small-Block Performance: 1-55788-323-8/HP1323
Mustang 5.0 Projects: 1-55788-275-4/HP1275
Mustang Performance (Engines): 1-55788-193-6/HP1193
Mustang Performance 2 (Chassis): 1-55788-202-9/HP1202
Mustang Perf. Chassis, Suspension, Driveline Tuning: 1-55788-387-4
Mustang Performance Engine Tuning: 1-55788-387-4/HP1387
Mustang Restoration Handbook ('64–'70): 0-89586-402-9/HP1402
Rebuild Big-Block Ford Engines: 0-89586-070-8/HP1070
Rebuild Ford V-8 Engines: 0-89586-036-8/HP1036
Rebuild Small-Block Ford Engines: 0-912656-89-1/HP189

MOPAR
Big-Block Mopar Performance: 1-55788-302-5/HP1302
How to Hot Rod Small-Block Mopar Engine Revised: 1-55788-405-6
How to Maintain & Repair Your Jeep: 1-55788-371-8/HP1371
How to Modify Your Jeep Chassis/Suspension for
  Offroad: 1-55788-424/HP1424
How to Modify Your Mopar Magnum V8: 1-55788-473-0/HP1473
How to Rebuild Your Mopar Magnum V8: 1-55788-431-5, HP1431
Rebuild Big-Block Mopar Engines: 1-55788-190-1/HP1190
Rebuild Small-Block Mopar Engines: 0-89586-128-3/HP1128
Torqueflite A-727 Transmission Handbook: 1-55788-399-8/HP1399

IMPORTS
Baja Bugs & Buggies: 0-89586-186-0/HP1186
Honda/Acura Engine Performance: 1-55788-384-X/HP1384
Honda/Acura Performance: 1-55788-384-X/HP1384
How to Hot Rod VW Engines: 0-912656-03-4/HP103
Porsche 911 Performance: 1-55788-489-7/HP489
Rebuild Air-Cooled VW Engines: 0-89586-225-5/HP1225
The VW Beetle: A History of The World's Most Popular Car:
  1-55788-421-8/HP1421
The Volkswagen Super Beetle Handbook: 1-55788-483-8/HP1483

HANDBOOKS
Automotive Detailing: 1-55788-288-6/HP1288
Auto Electrical Handbook: 0-89586-238-7/HP1238
Auto Math Handbook: 1-55788-020-4/HP1020
Automotive Paint Handbook: 1-55788-291-6/HP1291
Auto Upholstery & Interiors: 1-55788-265-7/HP1265
Car Builder's Handbook: 1-55788-278-9/HP1278
Classic Car Restorer's Handbook: 1-55788-194-4/HP1194
Engine Builder's Handbook: 1-55788-245-2/HP1245
Fiberglass & Composite Materials: 1-55788-239-8/HP1239
The Lowrider's Handbook: 1-55788-383-1/HP1383
Metal Fabricator's Handbook: 0-89586-870-9/HP1870
1001 High Performance Tech Tips: 1-55788-199-5/HP1199
1001 MORE High Performance Tech Tips: 1-55788-429-3/HP1429
Paint & Body Handbook: 1-55788-082-4/HP1082
Performance Ignition Systems: 1-55788-306-8/HP1306
Pro Paint & Body: 1-55788-394-7/HP1394
Sheet Metal Handbook: 0-89586-757-5/HP1757
Welder's Handbook: 1-55788-264-9/HP1264

INDUCTION
Holley 4150: 0-89586-047-3/HP1047
Holley Carbs, Manifolds & F.I.: 1-55788-052-2/HP1052
Rochester Carburetors: 0-89586-301-4/HP1301
Tuning Accel/DFI 6.0 Programmable F.I: 1-55788-413-7/HP1413
Turbochargers: 0-89586-135-6/HP1135
Street Turbocharging: 1-55788-488-9/HP1488
Weber Carburetors: 0-89586-377-4/HP1377

RACING & CHASSIS
Bracket Racing: 1-55788-266-5/HP1266
Chassis Engineering: 1-55788-055-7/HP1055
4Wheel&Off-Road's Chassis & Suspension: 1-55788-406-4/HP1406
How to Make Your Car Handle: 0-912656-46-8/HP146
How to Get Started in Stock Car Racing: 1-55788-468-4/HP1468
Race Car Engineering and Mechanics: 1-55788-366-1/HP1366
Stock Car Setup Secrets: 1-55788-401-3/HP1401

MOTORCYCLES
American V-Twin Engine Tech: 1-55788-455-2/HP1455

STREET RODS
How to Build a 1932 Ford Street Rod: 1557884781/HP1478
How to Build a 1933–34 Ford Street Rod: 1-55788-479-X/HP1479
How to Build a 1936–1940 Ford Street Rod: 1-55788-493-5/HP1493
Street Rodder magazine's Chassis & Suspension Handbook
  1-55788-346-7/HP1346
Street Rodder's Handbook, Rev.: 1-55788-409-9/HP1409

ORDER YOUR COPY TODAY!
All books can be purchased at your favorite retail or online bookstore (use ISBN number), or auto parts store (use HP part number). You can also order direct from HPBooks by calling toll-free at 800-788-6262, ext. 1.